ASSISTED VENTILATION

ASSISTED VENTILATION

Edited by

John Moxham, MD, FRCP
Professor of Thoracic Medicine,
King's College School of Medicine
and Dentistry, London

Articles reprinted from Thorax
Published by the British Medical Journal
Tavistock Square, London WC1H 9JR

British Library Cataloguing in Publication Data
Assisted Ventilation.
616.2

ISBN 0727903063

Front cover photograph: Courtesy of John Watney Picture Library

Printed in Great Britain by Eyre & Spottiswoode Ltd, London and Margate

Contents

Introduction

JOHN MOXHAM

This book consists of contributions from thoracic physicians, intensive care specialists, and anaesthetists. Its purpose is to provide comprehensive information about assisted ventilation, aimed at thoracic and general physicians rather than anaesthetists. Many physicians care for patients on the intensive care unit, and it is particularly appropriate for thoracic physicians to be involved in the management of such patients. If physicians, including those in training, are to feel at home on the intensive care unit, they must be familiar with all aspects of assisted ventilation.

Chapter 1 describes the techniques of assisted ventilation that are available on the intensive care unit, and seeks to unravel the jargon and explain the equipment. Too frequently, physicians on the unit do not know the details of ventilation technique. Ventilation cannot be isolated from the function of other organ systems and overall care of patients is not possible without an understanding of the process of positive pressure ventilation.

Chapter 2 details the indications for assisted ventilation; it is as important to appreciate the limitations of the technique as much as its possibilities. Hopefully this section will contribute to the debate that so frequently occurs between physician and anaesthetist about the merits of ventilating particular patients.

Ventilated patients on intensive care units are totally dependent on those caring for them, for the maintenance of all organ systems. Thus to care for the intubated patient is a wide and detailed task. Chapter 3 seeks to discuss the many and varied aspects of the general care of intubated patients.

For most ventilated patients on the intensive care unit, the withdrawal of assisted ventilation is straightforward, but in a minority, weaning presents great difficulties. Weaning has long been seen as an art, mainly the art of the anaesthetist. Such weaning difficulties often cause concern and frustration. Chapter 4 seeks to explain why some patients fail to wean, what can be done to facilitate the weaning process, and in general the chapter seeks to demystify this complex clinical situation.

For most patients assisted ventilation occurs on the intensive care unit, but, increasingly, selected patients are receiving ventilation within the community, most commonly to control the nocturnal hypoventilation associated with skeletal and neuromuscular disorders such as kyphoscoliosis. Non-invasive and novel techniques have revolutionised this field, and the indications for domiciliary ventilation and hospital-based ventilation outside the intensive care unit will probably broaden. The clinical indications for domiciliary ventilation and the range of techniques now available are discussed in chapters 5 and 6.

I hope that having worked through the indications for ventilation, the techniques available, the general care of intubated patients, the problems of weaning, and the question of domiciliary ventilation, the reader will feel that there is not much in the field of assisted ventilation that is likely to cause him or her too much anxiety.

1

Artificial ventilation: history, equipment, and techniques

J D YOUNG, M K SYKES

A brief history of artificial ventilators

An artificial ventilator is essentially a device that replaces or augments the function of the inspiratory muscles, providing the necessary energy to ensure a flow of gas into the alveoli during inspiration. When this support is removed gas is expelled as the lung and chest wall recoil to their original volume; exhalation is a passive process. In the earliest reports of artificial ventilation this energy was provided by the respiratory muscles of another person, as expired air resuscitation. Baker[1] has traced references to expired air resuscitation in the newborn as far back as 1472, and in adults there is a report of an asphyxiated miner being revived with mouth to mouth resuscitation in 1744. In the eighteenth century artificial ventilation became the accepted first line treatment for drowning victims, though the use of bellows replaced mouth to mouth resuscitation.[2] Automatic artificial ventilators that did not require a human as a power source took another 150 years to appear, being first suggested by Fell[3] and then made available commercially by Draeger[4] in 1907. These were still devices for resuscitation, for the Draeger company at that time was noted for its mine rescue apparatus.

The introduction of artificial ventilators into anaesthesia proceeded slowly until surgeons ventured into the chest. During thoracic surgery on spontaneously breathing patients the inevitable pneumothoraces and mediastinal shift ("mediastinal flap") that occurred as the pleural cavity was opened caused a high mortality, which was substantially reduced when positive pressure ventilation was used. A further boost to the development of automatic artificial ventilators occurred in 1952, when a catastrophic poliomyelitis epidemic struck Denmark. There was a very high incidence of bulbar lesions and 316 out of 866 patients with paralysis admitted over a period of 19 weeks required postural drainage, tracheostomy, or respiratory support. By using tracheostomy and manual positive pressure ventilation the Danish physicians reduced the mortality from poliomyelitis from 80% at the beginning of the epidemic to 23% at the end. The artificial ventilation was entirely by hand, a total of 1400 university students working in shifts to keep the patients ventilated. The fear that another epidemic might afflict Europe expedited research into powered mechanical ventilators.

Classification of artificial ventilators

The lungs can be artificially ventilated either by reducing the ambient pressure around the thorax (negative pressure ventilation) or by increasing the pressure within the airways (positive pressure ventilation). Negative pressure ventilators use a rigid chamber that encloses either the thorax (cuirass) or the whole body below the neck (tank respirator or "iron lung"). The pressure in the chamber is reduced cyclically by means of a large volume displacement pump, thus causing the lungs to expand and contract. Negative pressure ventilation is fully discussed in chapter 5 of this series. These ventilators were used extensively for poliomyelitis victims, and are still in use for long term respiratory support or overnight support for patients with respiratory muscle weakness. Tank ventilators occupy much space, access to the patient is poor, and the neck seal can create problems. They are not suitable for use in general intensive care units. There has been a recent revival in interest in negative pressure ventilators in paediatric intensive care units to avoid the need for endotracheal intubation.[5]

The basic classification of positive pressure artificial ventilators was first proposed by Mapleson.[6] Artificial ventilators are devices that control inspiration; expiration is usually passive, and so the classification is based on the mechanism of gas delivery during inspiration. There are two types of machine. A flow generator produces a known pattern of gas flow during inspiration, and the lungs fill at a rate entirely controlled by the ventilator and independent of any effect of lung mechanics. A pressure generator produces a preset pressure in the airway and the rate of lung inflation depends not only on the pressure generated by the ventilator but

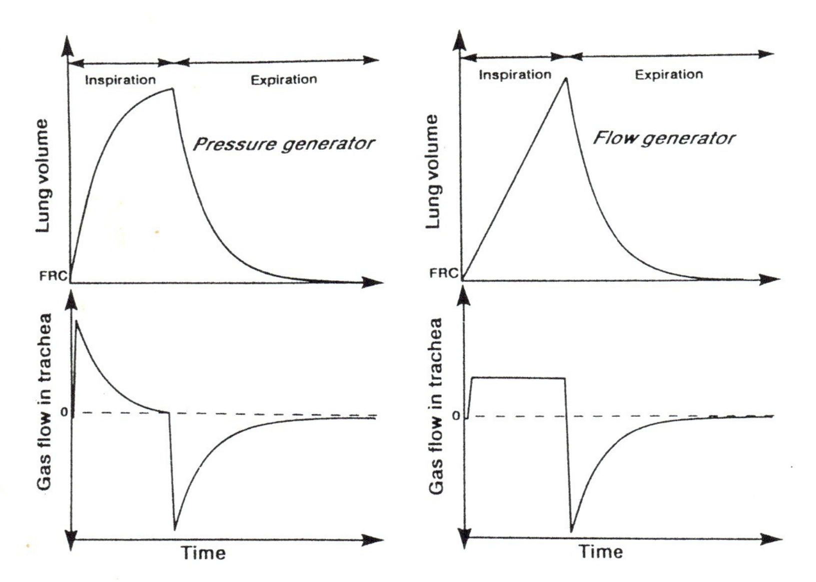

Figure 1 Lung volume and gas flow during one respiratory cycle with constant pressure and constant flow generator ventilators.

also on the respiratory resistance and compliance, which determine the time constant of the lungs. With a square wave pattern of pressure the lungs fill in an exponential fashion. The effects of the two types of ventilator on inspiratory flow and lung volume are shown in figure 1.

In general, the flow generator ventilator is used for adults and the pressure generator ventilator for children, or adults when control of peak airway pressure is important. The pressure generator is particularly useful in children where uncuffed endotracheal tubes are used and there is a leak of gas around the endotracheal tube during inspiration. A pressure generator tends to compensate for this leak by increasing the flow into the airway, whereas with a flow generator a proportion of the tidal volume is lost.

A ventilator must also have a mechanism to cause it to cycle between inspiration and expiration. Ventilators are usually subclassified as time cycled, pressure cycled, or volume cycled machines. Time cycled ventilators switch between inspiration and expiration after a preset time interval, pressure cycled ventilators switch when a preset airway pressure threshold has been reached, and volume cycled ventilators cycle when a preset tidal volume has been delivered. Cycling from expiration to inspiration is usually effected by a timing mechanism or by a patient triggering device that senses the subatmospheric pressure or the flow generated in the inspiratory tube by the patient's inspiratory effort.

This traditional classification was devised during a period when ventilators were totally mechanical, and driven either by compressed gas or by an electrically powered piston or bellows. By 1954, however, Donald had used an electronic trigger that initiated inspiration and in 1958 an electronic timing device was incorporated in the Barnet ventilator.[7] In 1971 the Siemens-Elema company introduced the Servo 900 ventilator, which combined a simple pneumatic system with a sophisticated electronic measuring and control unit.[8] Gas flow to and from the patient's lungs was controlled by a pair of scissor valves and monitored with pressure and flow sensors. The control unit adjusted the scissor valves to ensure that the flow patterns measured by the sensors corresponded to those selected by the operator. This method of control is termed a servo or feedback system and is very flexible. Potentially one machine could mimic all the previously described categories of ventilator.

Modern intensive care unit ventilators

Most modern intensive care unit ventilators use similar technology. A simplified diagram of such a ventilator is shown in figure 2. The respiratory gas is held in a pressurised reservoir and delivered to the patient via an inspiratory valve. The inspiratory valve and hence the inspiratory flow is controlled by the electronic control unit. The airway pressure and flow of gas into the patient are monitored by the pressure and inspiratory flow sensors. The expiratory flow can also be monitored to check for leaks and disconnection of the patient from the ventilator. This design enables the ventilator to be used as either a flow or a pressure generator. For example, if the operator selects a constant flow during inspiration the ventilator will open the inspiratory valve until the flow sensor measures the required flow. If the inspiratory flow decreases the inspiratory valve will be opened further to compensate, and vice versa. If the operator wishes the ventilator to act as a constant pressure generator, the ventilator will open the inspiratory valve until the pressure sensor indicates that the desired pressure has been reached. The ventilator will then maintain the airway pressure at the desired level by opening or

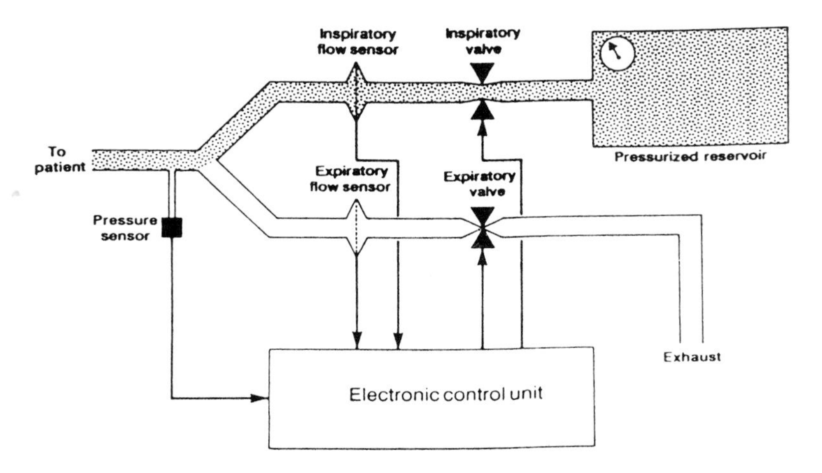

Figure 2 A block diagram of a modern ventilator. The shaded area indicates the gas path during inspiration.

closing the inspiratory valve. The bulk of a modern ventilator is now the electronic unit and the pneumatic units are often very compact. Figure 3 is a photograph of a Servo 900C ventilator. The pneumatic systems are housed in the top section and the bottom section contains only electronics.

The use of electronic feedback systems in ventilators has many advantages. The moving parts within the ventilator are kept to a minimum, sophisticated alarm systems are possible, the ventilators can be made physically small, and maintenance and repair are simple. If a new mode of ventilation appears to be useful clinically it can often be added to an existing machine by altering the software in the electronic unit—indeed, our ability to produce new modes of artificial ventilation far exceeds our ability to test them clinically.

Interactions between patients and ventilators

As early as 1929 it was observed that a patient who "fought" the ventilator was difficult to manage and suffered complications.[9] To minimise these problems many ventilated patients are sedated, and in some cases paralysed with neuromuscular blocking agents (for example, pancuronium). In the early 1980s paralysis of ventilated patients was very common in British intensive care units,[10 11] and sedation was used to dissociate the patients from their surroundings. Over the last decade the amount of sedative and paralysing

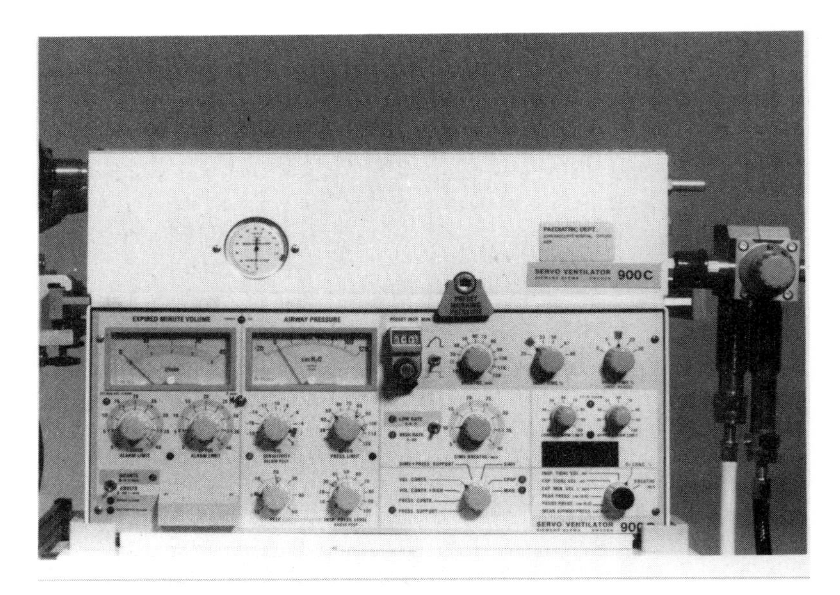

Figure 3 A Siemens Servo 900 C ventilator.

drugs given to intensive care patients has been reduced considerably.

Paralysis of ventilated patients is not without risks.[12] The paralysed patient is unable to make any spontaneous respiratory effort, so a disconnection from the ventilator is rapidly fatal. The patient may be aware of his or her surroundings but is unable to communicate, and the risk of pulmonary embolus is probably increased. Even sedative drugs are not without risk,[13] as shown by the substantial increase in mortality in trauma patients given etomidate infusions. This was subsequently shown to be caused by suppression of adrenal activity by the drug.

Only a few clinicians now favour deep sedation and paralysis,[14] the preferred approach being a comfortable patient who can be easily aroused and can communicate with the staff. To achieve this goal ventilators have to be more flexible and adapted to the needs of the patient. In its simplest form this is achieved by inserting a valve into the ventilator tubing, which allows the patient to breathe spontaneously from another gas source between the breaths delivered by the ventilator. This mode of ventilation is termed intermittent mandatory ventilation (IMV) and was initially used as a technique to wean patients from mechanical ventilation to spontaneous respiration. By 1987, however, over 70% of American intensive care units were using intermittent mandatory ventilation as their primary mode of ventilatory support.[15] It is not a truly interactive mode of ventilation—the ventilator continues to deliver the preset pattern of ventilation regardless of the patient's respiratory effort. The advent of microprocessor controlled ventilators has produced several "smart" modes of ventilation, where the ventilator adjusts the ventilatory support it supplies according to the patient's respiratory efforts. The commonest modes used are synchronised intermittent mandatory ventilation, mandatory minute ventilation, and inspiratory pressure support.

Ventilatory strategies in common use

INTERMITTENT POSITIVE PRESSURE VENTILATION

Each year millions of patients are ventilated during anaesthesia for surgical procedures that require muscular relaxation. Nearly all of these patients are ventilated with basic intermittent positive pressure ventilation by means of a simple mechanical ventilator. For these short periods of respiratory support these ventilators are perfectly adequate, for as the patient is both paralysed and anaesthetised there is no need for interactive ventilators. Weaning is accomplished by reversing the muscle relaxation and lightening the anaesthetic. In intensive care units this mode of ventilation is used when the patient has to be heavily sedated and paralysed to treat the primary condition (for example, tetanus) or when the patient is unable to make any respiratory movement (for example, severe Guillain-Barré syndrome). Intermittent positive pressure ventilation can be achieved with almost any artificial ventilator as no special refinements are needed.

INTERMITTENT MANDATORY VENTILATION

As outlined above, intermittent mandatory ventilation describes a mode of mechanical ventilation that allows the patient to breathe spontaneously through the ventilator circuit. At predetermined intervals a positive pressure breath is provided by the ventilator totally independently of the patient's spontaneous ventilatory pattern. The system is a simple mechanical device that can be fitted to almost any ventilator, though care has to be taken to ensure that the oxygen concentration is the same in the spontaneous breaths as

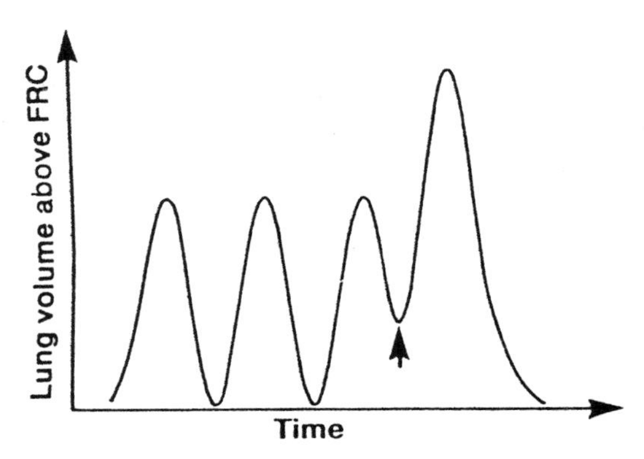

Figure 4 "Stacking" of breaths during intermittent mandatory ventilation. The arrow indicates where a mandatory breath starts before the patient has fully expired.

during the breaths delivered by the ventilator and that the inspired gas is appropriately humidified.[16] Though simple it is not without disadvantages. There is no synchronisation between the patient's respiratory efforts and the ventilator, and this can lead to "stacking" of breaths and high airway pressures. Figure 4 shows a plot of lung volume against time for a patient on intermittent mandatory ventilation. The patient can take a spontaneous breath, and before he has fully exhaled the ventilator delivers a breath. A very large tidal volume with high peak airway pressure results. To overcome this problem ventilators were developed that were able to detect spontaneous breaths taken by the patient.

SYNCHRONISED INTERMITTENT MANDATORY VENTILATION

To avoid "stacking" the ventilator must be able to sense that the patient has taken a breath, and then avoid delivering a mandatory breath during the period of the spontaneous breath. This level of sophistication was not easily achieved before electronic control was used in artificial ventilators. One of the first machines to have a synchronised intermittent mandatory ventilation facility was the Servo 900B, a refinement of the original Servo 900. This machine dispensed with external intermittent mandatory ventilation valves and used its internal valves and sensors to provide synchronised intermittent mandatory ventilation. Any spontaneous inspiratory activity by the patient was sensed by the pressure sensor and then the expiratory valve was closed and the inspiratory valve opened. The flow of gas through the inspiratory valve was matched to the patient's inspiratory flow rate by opening and closing the inspiratory valve. This allowed the patient to breathe spontaneously through the ventilator. When a mandatory breath was due the ventilator would wait until the patient began to inspire and then deliver the mandatory breath, synchronising the mandatory breath with the patient's own inspiratory effort.

All ventilators with synchronised intermittent mandatory ventilation systems use some form of synchronisation between patient and machine, the exact details varying between machines. All synchronised intermittent mandatory ventilation machines are not equally effective, for the time taken for the inspiratory valve to open after the pressure sensor has been triggered varies between machines. During the period between the beginning of the patient's inspiration and the opening of the inspiratory valve the patient inspires against a closed valve, increasing the work of breathing.[17] Machines with a short lag between the beginning of an inspiratory effort and the opening of the inspiratory valve are preferred. Although subjectively patients usually appear more comfortable with synchronised intermittent mandatory ventilation than intermittent positive pressure ventilation or intermittent mandatory ventilation, there is no evidence that it reduces morbidity or mortality in intensive care units.

MANDATORY MINUTE VENTILATION

The synchronised intermittent mandatory ventilation systems described add a preset number of breaths of a preset volume to the patient's own respiratory efforts. If the patient's spontaneous breaths are providing most of the required minute ventilation, the mandatory breaths may be too large or too frequent. Alternatively, if the patient's own respiratory efforts diminish the mandatory breaths may not be enough to provide an adequate minute ventilation. Ideally the ventilator should monitor exhaled volumes and "top up" the patient's respiratory efforts if needed. This is mandatory minute ventilation. In this mode of ventilation a preset minute volume is selected. If the patient's own respiratory efforts produce a minute ventilation equal to or greater than the preset value the ventilator is not activated. If the preset minute volume is not achieved the ventilator gives some assistance to the patient, to bring the minute ventilation back to the target value. This system works well if the patient's respiratory rate is relatively normal. If, however, the patient has rapid, shallow breaths the efficiency of breathing will be reduced and the mandatory volume may become inadequate.

The method used to provide ventilatory assistance depends on the design of the ventilator. The Ohmeda CPU1 and Engstrom Erica will provide synchronised mandatory breaths of a preset tidal volume at an increasing frequency as the patient's own respiratory efforts diminish. The Hamilton Veolar uses a different approach and provides increasing inspiratory pressure support (see below) as the patient's own efforts decrease. Whichever method is used, the machines ensure that the patient always receives a predetermined minute volume.

INSPIRATORY PRESSURE SUPPORT

The most recent mode of ventilation to be widely used is inspiratory pressure support. As the patient initiates a breath the ventilator raises the airway pressure to a preset value. The positive airway pressure provides some of the energy needed to expand the lung and the efforts of the patient provide the rest. At the end of inspiration the positive airway pressure is removed to allow unimpeded expiration. By selection of an appropriate level of airway pressure patients can be given only the respiratory assistance they actually require. The patient also determines his or her own respiratory rate. This is a relatively new mode of ventilation and has not yet been fully assessed. This form of ventilation differs from the assist mode commonly used in the 1960s. In the assist mode the magnitude of the breath was preset by the adjustment of controls on the ventilator and the breath was initiated whenever the patient trigger was activated. In inspiratory pressure support a constant pressure is applied after the patient triggers the ventilator so that the patient can determine the flow pattern and size of breath. Inspiration is terminated when the inspiratory flow ceases.

POSITIVE END EXPIRATORY PRESSURE

Patients who are anaesthetised or comatose or who have recently undergone abdominal surgery have a reduced functional residual capacity. The reduction may be sufficient to cause airway closure in dependent areas of the lung before the end of expiration, leading to underventilation of these areas and an increase in intrapulmonary shunt.[18] Widespread alveolar collapse also occurs in patients with the adult respiratory distress syndrome and results in severe hypoxaemia. The functional residual capacity can be maintained by leaving a constant standing pressure on the lungs, keeping them slightly inflated even at the end of expiration. This is positive end expiratory pressure or PEEP. The use of PEEP, and how best to determine its optimum level, has been the subject of debate for many years. On the positive side, it causes an increase in arterial oxygen tension for the same inspired oxygen tension. On the negative side, the constantly raised intrathoracic pressure causes a diminution in venous return to the heart, a decrease in cardiac output, and an increased risk of pneumothorax. The controversy may now be resolved, for a recent study showed that mortality increased with the aggressive application of PEEP and that the minimum level which kept arterial oxygenation within acceptable limits was "best PEEP."[19]

UNUSUAL MODES OF VENTILATION

There are several modes of ventilation that are still being evaluated clinically. In high frequency jet ventilation brief, frequently repeated pulses of gas (up to 300 a minute) are directed from a high pressure nozzle down the airway. The tidal volumes delivered are small, airway pressures are low, and the patient can breathe spontaneously while being ventilated.[20] High frequency oscillation techniques oscillate gas in the airways using a piston pump or diaphragm pump. The tidal volumes used are very small, much less than the anatomical dead space, so gas movement occurs by mechanisms other than tidal exchange. Because of the large size of oscillators needed for adults this technique has been confined to use in infants. There are also two extracorporeal systems in clinical use. In adults with acute respiratory failure extracorporeal membrane oxygenation was abandoned after a large trial showed that it did not reduce mortality.[21] In infants with neonatal respiratory distress, however, the method seems to be very effective. A remarkable reduction in mortality from this condition has been reported in America when extracorporeal membrane oxygenation has been instigated.[22] In adults a new approach has been pioneered by Gattinoni in Italy. He started with the hypothesis that artificial ventilation itself may exacerbate acute respiratory failure. Tidal ventilation is required only to remove carbon dioxide, and oxygenation can be maintained in apnoeic patients if the carbon dioxide is removed by an extracorporeal circuit. By removing carbon dioxide with an extracorporeal artificial lung and reducing ventilation to the minimum he has claimed to reduce mortality from acute respiratory failure.[23]

Conclusion

The distinction between a ventilated patient and a spontaneously breathing patient is becoming increasingly blurred as more sophisticated means of respiratory support are devised. In many cases "respiratory assistance" may be a more appropriate term than "artificial ventilation." The efficacy of many of these ventilatory strategies has not been properly assessed, in terms either of acceptability to the patient or of the effect on morbidity and mortality. It is still true, however, that "the type of ventilator used, provided that its design is satisfactory, is of less importance than the experience of the person using it."[24]

1 Baker AB. Early attempts at expired air respiration, intubation and manual ventilation. In: Atkinson RS, Boulton TB, eds. *The history of anaesthesia.* London: Royal Society of Medicine, 1987:372–4.
2 Kite C. *An essay on the recovery of the apparently dead.* London: Dilly, 1788.
3 Goerig M, Filos K, Ayisi KW. George Edward Fell and the development of respiratory machines. In: Atkinson RS, Boulton TB, eds. *The history of anaesthesia.* London: Royal Society of Medicine, 1987:386–93.
4 Rendell-Baker L, Pettis JL. The development of positive pressure ventilators. In: Atkinson RS, Boulton TB, eds. *The history of anaesthesia.* London: Royal Society of Medicine, 1987: 402–21.
5 Samuals MP, Southall DP. Negative extrathoracic pressure in the treatment of respiratory failure in infants and young children. *Br Med J* 1989;**299**:1253–7.
6 Mapleson WW. The effect of lung characteristics on the functioning of artifical ventilators. *Anaesthesia* 1962;**31**:300–14.
7 Mushin WW, Rendell-Baker L, Thompson PW, Mapleson WW. Automatic ventilation of the lungs. Oxford: Blackwells, 1980:312–30.
8 Ingelstedt S, Johnson B, Nordstrom L, Olsson S-G. A servo-controlled ventilator measuring expired minute volume, airway flow and pressure. *Acta Anaesthesiol Scand* 1972;suppl **47**:9–28.
9 Drinker P, McKhann CF. The use of a new apparatus for prolonged administration of artificial respiration. *JAMA* 1929;**92**:1658–61.
10 Miller-Jones, CMH, Williams JH. Sedation for ventilation. A retrospective study of fifty patients. *Anaesthesia* 1980;**35**:1104–7.
11 Merriman HM. The techniques used to sedate ventilated patients. *Intens Care Med* 1981; 7:217–24.
12 Willatts SM. Paralysis for ventilated patients? Yes or No? *Intens Care Med* 1985;**11**:2–4.
13 Watt I, Ledingham I. McA. Mortality amongst multiple trauma patients admitted to an intensive therapy unit. *Anaesthesia* 1984;**39**:973–81.
14 Bion JF, Ledingham I. McA. Sedation in intensive care—a postal survey. *Intens Care Med* 1987;**13**:215–6.
15 Venus B, Smith RA, Mathru M. National survey of methods and criteria used for weaning from mechanical ventilation. *Crit Care Med* 1987;**15**:530–3.
16 Pybus DA, Kerr JH. A simple system for adminstering intermittent mandatory ventilation (IMV) with the Oxford ventilator. *Br J Anaesth* 1978;**50**:271–4.
17 Lemaire LF. Weaning from mechanical ventilation. In: Ledingham I, McA, ed. *Recent advances in critical care medicine 3.* Edinburgh: Churchill Livingstone, 1988:15–30.
18 Hedenstierna G. Causes of gas exchange impairment during general anaesthesia. *Eur J Anaesthesiol* 1988;**5**:221–31.
19 Carroll GC, Tuman KJ, Braverman B, *et al.* Minimal positive end-expiratory pressure (PEEP) may be best PEEP. *Chest* 1988;**93**:1020–5.

20 Smith BE, Hanning CD. Advances in respiratory support. *Br J Anaesth* 1986;**58**:138–50.
21 Zapol WM, Snider MT, Hill JD, *et al.* Extracorporeal membrane oxygenation in severe acute respiratory failure. *JAMA* 1979;**242**:2193–6.
22 Toomasian JM, Snedcor SM, Cornell RG, Cilley RE, Bartlett RH. National experience with extracorporeal membrane oxygenation for newborn respiratory failure. *Trans Am Soc Artif Intern Organs* 1988;**34**:140–7.
23 Gattinoni L, Pesenti A, Mascheroni D, *et al.* Low-frequency positive-pressure ventilation with extra-corporeal CO_2 removal in severe acute respiratory failure. *JAMA* 1986; **256**:881–6.
24 Sykes MK, McNicol MW, Campbell EJM. *Respiratory failure.* Oxford: Blackwells, 1976:212.

2

Indications for mechanical ventilation

JOSÉ PONTE

Mechanical ventilation comprises all types of artificial ventilation in which a mechanical device is used to replace or aid the work normally carried out by the ventilatory muscles. It has been used to treat ventilatory failure since the portable "iron lung" was introduced by Drinker and Shaw in 1929,[1] but only during the Copenhagen poliomyelitis epidemic in 1952 were the skills of anaesthetists and physicians brought together in a major breakthrough that established the lifesaving value and simplicity of intermittent positive pressure ventilation.[2,3] This was an important landmark in the treatment of acute respiratory failure and soon afterwards the benefits of intermittent positive pressure ventilation in the postoperative period were also proved.[4] Unless otherwise stated, this chapter refers to mechanical ventilation as any form of intermittent positive pressure ventilation, applied through an endotracheal tube, with or without positive end expiratory pressure or allowance for spontaneous breaths.

Mechanical ventilation is indicated where established or impending respiratory failure exists, defined as the inability of the breathing apparatus to maintain normal gas exchange. Respiratory failure may be predominantly due to failure of oxygenation (type I) or to an inability to eliminate carbon dioxide (type II or "ventilatory" failure). Type I is usually associated with lung parenchymal disease, alveolar collapse, or an increase in lung water. Type II is generally associated with a lack of ventilatory drive, musculoskeletal disease, or neuromuscular blockade. In principle, mechanical ventilation is indicated predominantly for ventilatory failure (type II).

The vast majority of patients receiving mechanical ventilation do not have pulmonary disease. This is the group undergoing major surgery under general anaesthesia with the aid of neuromuscular blocking drugs. Outside the operating theatre most patients receive mechanical ventilation in the intensive therapy unit. Within non-specialised intensive therapy units (table 1) over half of all ventilated patients have had cardiac, aortic, or other major surgery, rarely needing intermittent positive pressure ventilation for more than 24 hours. The other major groups requiring ventilation are patients with head or chest trauma (<15%) and various forms of poisoning (<8%) and those who are critically ill with severe primary respiratory (<15%) or cardiac (<3%) disease.[5-7] A very small group of patients receive mechanical ventilation at home or in specialised institutions; these will be considered separately in chapters 5 and 6.

A comprehensive list of indications for mechanical ventilation

Table 1 Underlying pathological conditions of patients admitted to non-specialised intensive therapy units in three surveys

	No of patients	*% of total*
A Pontoppidan et al[5] (1972): patients admitted to intensive therapy units over nine consecutive years (1961–70) requiring mechanical ventilation; mortality 35–40% in the first five years, then 10–20%		
Cardiothoracic surgery	239	25·9
Neurosurgery including head trauma	69	7·5
Other surgery including trauma	243	26·3
Neuromuscular disease	71	7·7
Poisoning	73	7·9
Primary respiratory disease	121	13·1
Cardiac failure	12	1·3
Other	95	10·3
Total	923	
B Petty et al[6] (1975): 1877 patients admitted to intensive therapy units over 10 consecutive years (1964–74) who received ventilation for more than 24 hours; survival rate was 75·2%		
Postoperative	1009	53·8
Neurosurgical	115	6·1
Trauma	168	8·9
Poisoning	145	7·7
Primary respiratory disease	250	13·3
Total	1877	
C Knaus et al[7] (1989): 3884 consecutive patients admitted to intensive therapy units in 12 United States hospitals, of whom 1886 (49%) received mechanical ventilation		
Postoperative	1315	70
Others	571	30

Table 2 List of possible indications for mechanical ventilation

List of possible indications
Routine anaesthesia
Cardiothoracic and abdominal surgery and neurosurgery
Prolonged surgery and surgery requiring prone position
Surgery in frail patients or those with cardiac disease
Clinical investigations (radiology, tissue biopsy) requiring temporary immobility
Postoperative management
Major surgery of the heart or the great vessels
Abdominal distention, debility, or electrolyte imbalance
Pre-existing lung disease, respiratory muscle weakness, kyphoscoliosis, myasthenia gravis, morbid obesity
Respiratory disease (parenchymal or airway)
Pneumonia, asthma, lung contusion
Acute exacerbation of chronic bronchitis, emphysema
Adult respiratory distress syndrome, hyaline membrane disease, cystic fibrosis
Chest wall disease
Trauma with flail segment, ruptured diaphragm
Chest wall burns, kyphoscoliosis
Neuromuscular disease
Polyneuritis, Guillain-Barré disease, Lambert-Eaton disease
Myasthenia gravis, myopathies, paralysing poisons
Central nervous system impairment
Drug overdose: narcotics, anaesthetics, barbiturates
Trauma, meningoencephalitis, tumours, infarction
Brain oedema, raised intracranial pressure
Intracranial bleed, status epilepticus, tetanus, rabies
Central hypoventilation
Cardiovascular disease
Cardiac arrest, severe shock—sepsis or other causes
Left ventricular failure—pulmonary oedema
Neonatal conditions
Severe prematurity
Severe bronchopulmonary dysplasia
Central hypoventilation syndrome
Increased metabolism and carbon dioxide production precipitating ventilatory failure in patients with pre-existing disease
Organ donation

appears in table 2. Indications for mechanical ventilation in anaesthesia, after surgery, in neonates and in organ donation are outside the scope of this review. Mechanical ventilation should be used only when it is strictly necessary as there are many inherent risks.[8] Indeed, ventilation may unnecessarily prolong the distress of terminal disease and the benefits of its use should therefore be carefully weighed against the disadvantages. The basic "recipe" for

Table 3 Basic "recipe" for setting up mechanical ventilation in an adult patient without pulmonary disease and with a normal metabolic rate

AIRWAY
- Access via oral or nasal cuffed endotracheal tube or cuffed tracheostomy tube

VENTILATOR
- Set tidal volume (VT) at 6 ml/kg body weight
- Set respiratory rate (RR) at 12–14 breaths/min; minute volume (= VT × RR) should be 80–90 ml/kg body weight
- Set ratio of inspiratory:expiratory time to 1:3; peak inflation pressure should not exceed 30 cm H_2O
- Provide humidification of inspired gas mixture
- Set oxygen concentration at 30–60%
- Set alarms for
  - ventilator disconnection
  - peak inspiratory pressure > 30 cm H_2O
  - inspired oxygen concentration 25%–60%

PATIENT
- Ensure analgesia and sedation: mandatory if patient is paralysed (neuromuscular blockers only when strictly necessary—that is, tetanus, head injuries)—use opiates and/or benzodiazepines
- Monitor effects of intermittent positive pressure ventilation on circulation and gastric distension
- Check blood gas tensions regularly (2–4 hourly) and after changing any of the ventilator settings:
  - adjust minute volume according to arterial carbon dioxide tension
  - adjust FIO_2 and positive end expiratory pressure according to arterial oxygen tension
- Institute basic nursing care for the unconscious patient:
  - regular routine observations, turning on bed, mouth wash
  - regular check for bilateral breath sounds and expansion of both lungs—risk of endobronchial intubation, pneumothorax, accumulation of secretions
  - regular check for state of consciousness, need of pain relief, and sedation
- Chest radiograph on alternate days to check for:
  - position of endotracheal tube
  - pleural effusion
  - alveolar collapse or consolidation

setting up intermittent positive pressure ventilation, in a patient without lung disease, appears in table 3.

Benefits of mechanical ventilation

The principal benefit of mechanical ventilation is the control gained over the airway and over the work of breathing. The ventilator replaces the work of exhausted or temporarily inadequate respiratory muscles. The ability to remove secretions from upper airways (by simple suction or aided by fibreoptic bronchoscopy) may be advantageous and additional measures, such as positive end expiratory pressure or effective aerosol delivery, may be instituted in certain patients. Ventilation also allows large doses of narcotic analgesics or neuromuscular blocking agents to be used where clinically indicated (in tetanus, for example). Table 2 lists clinical circumstances in which mechanical ventilation may be of benefit.

Risks and side effects of mechanical ventilation

THE RISKS

Some dangers of mechanical ventilation apply to all patients. Effective, long term intermittent positive pressure ventilation cannot be established without securing a sealed connection with the airway via an endotracheal or tracheostomy tube; the insertion of this tube, however, requires either general or local anaesthesia with its attendant risks.

Anaesthesia

The risks of the anaesthesia needed for endotracheal intubation include myocardial depression caused by general or local anaesthetics; aspiration of gastric contents; a further fall in arterial oxygen tension (PaO_2), especially if intubation is difficult; an idiosyncratic reaction to anaesthetic drugs; and reflex worsening of bronchoconstriction after tracheal intubation or suction of secretions. These risks are not substantially reduced if a topical local anaesthetic is used before intubation of the trachea.

Sedation and paralysis

Intermittent positive pressure ventilation through a nasal or an orotracheal tube is poorly tolerated without sedation and often requires paralysing drugs. The ideal sedative should be very short acting and be given by constant intravenous infusion, and should have minimal side effects, especially on the circulation. None of the available sedatives is devoid of side effects: the opiates are complicated by tolerance and paralysis of the gut (with consequent delay in absorption) and their prolonged respiratory depressant effects delay weaning from the ventilator. Barbiturates and chlormethiazole present similar problems and also cause myocardial depression. The benzodiazepines often require increasing dosage because of tolerance and this causes prolonged depressant effects, lasting for days after the last dose, on the central nervous system. Of the established anaesthetics, only two recently introduced drugs, currently being tested for long term sedation, have potential as sedatives for the intensive therapy unit—propofol,[9] an intravenous anaesthetic, and isoflurane,[10] a volatile inhalation anaesthetic. They are both short acting agents, with no cumulative effects, and are relatively free of cardiovascular and respiratory effects at sedative doses. There is a wide choice of suitable neuromuscular blocking drugs: pancuronium, vecuronium, and atracurium have minimal side effects and the last two are sufficiently short acting to allow rapid regulation of the state of paralysis. All intensive therapy unit staff should be aware that neuromuscular blocking agents have no sedative effects and that patients may be awake and paralysed if sedation is not prescribed. Another danger of paralysis is the inability of the patient to make spontaneous breathing efforts should there be an accidental ventilator disconnection.

Equipment failure

The risks of equipment failure include accidental disconnection of the ventilator, undetected leaks or malfunction of the endotracheal tube, all leading to alveolar hypoventilation; barotrauma to the lungs if high inflation pressures are applied to the airway in error, leading to pneumothorax or subcutaneous emphysema; tracheal burns if heated humidifiers are used; oxygen toxicity if F_{IO_2} is above 0·6 for a prolonged period.

Hyperinflation

Hyperinflation not associated with equipment failure occurs in patients with severe acute bronchospasm or chronic airflow limitation with increased functional residual capacity. Intermittent positive pressure ventilation may lead to very high intrathoracic pressures, with adverse effects on cardiac output and increased risk of pneumothorax.

Risks from the cardiovascular effects

Mean intrathoracic pressure is raised by intermittent positive pressure ventilation, especially where positive end expiratory pressure is used, causing a fall in cardiac output. These effects are unimportant in the relatively fit patient[11] undergoing elective surgery but may not be tolerated in the severely ill patient with hypovolaemia or with increased airways resistance.

CARDIOVASCULAR EFFECTS OF INTERMITTENT POSITIVE PRESSURE VENTILATION

Not all the effects of intermittent positive pressure ventilation on the cardiovascular system are adverse. They result from the rise in intrathoracic pressure, especially if positive end expiratory pressure is used,[12] and are mediated through direct mechanical interference with the heart,[13] through indirect reflexes of the autonomic nervous system, and through hormone release or changes in blood gases. The predominant direct adverse effects of intermittent positive pressure ventilation on the right heart are a reduction in venous return (preload) and an increase in pulmonary vascular resistance (afterload). The direct effects on the left heart are less pronounced and less well established, the widely held view being that both pulmonary venous return (preload) and afterload decrease. This effect on left ventricular afterload is due to a fall in ventricular transmural pressure because of the increase in intrathoracic pressure (this also applies to the right ventricle).[14] This mechanism provides a form of "assistance" to ventricular work, which may be beneficial in cardiac failure.[13,15] The reflex responses are complex, depending on multiple neural and chemical feedback loops.[16] The neural reflexes are mediated initially by the vagus nerve; stronger reflexes include the whole sympathetic system,[17] affecting vascular resistances and circulating catecholamines. The reflexes originate from lung and atrial stretch receptors and from the arterial baroreceptors and chemoreceptors (the latter only if arterial carbon dioxide tension ($Paco_2$) falls or oxygen tension (Pao_2) rises in response to intermittent positive pressure ventilation). The humoral reflex response to intermittent positive pressure ventilation includes an increase in antidiuretic hormone and renin-angiotensin and a decrease in atrial natriuretic peptide, which may be partly responsible for the sodium retention seen in ventilated patients[18]; the changes in catecholamines are partly mediated by $Paco_2$ changes. The pattern of circulatory changes is variable: the predominant effect is a decrease in both cardiac output (typically by 25%) and arterial blood pressure, an increase in heart rate, and a slight increase in systemic vascular resistance; right and left atrial pressures increase in relation to atmospheric pressure (transmural pressures *decrease*). This pattern is often modified by blood gas changes, such as the improvement in oxygenation usually associated with intermittent positive pressure ventilation; large changes in $Paco_2$ cause the largest changes in circulatory variables because of the powerful effects of carbon dioxide on the sympathetic system. In addition to cardiovascular effects, the increased intrathoracic pressure also hinders the venous drainage from the head, which may be important when intracranial pressure is raised.

Criteria for initiation of mechanical ventilation

Despite the risks listed above, patients developing acute severe respiratory failure secondary to reversible conditions should be ventilated as a life saving measure. Examples are pneumonia, asthma, neuromuscular syndromes, head or chest trauma, pulmonary oedema secondary to heart failure, poisoning, and septic shock. In nearly all cases the indications for mechanical ventilation are clearcut and ventilation is needed for only a short time (under 48 hours); there are circumstances, however, where the indications are less clear.[19 20]

Firstly, it is important, but may be difficult, to decide whether mechanical ventilation is the most appropriate treatment for the patient. In the adult respiratory distress syndrome, for example, it is debatable whether mechanical ventilation with positive end expiratory pressure should be started before the onset of type II respiratory failure. Although this is routine in many centres, the available evidence suggests that early institution of intermittent positive pressure ventilation does not alter the course or final outcome.[21] In general, intermittent positive pressure ventilation without respiratory failure is rarely indicated except after major surgery or in the management of raised intracranial pressure.

A second circumstance in which the decision to ventilate is difficult is in acute respiratory failure associated with terminal malignancy, advanced AIDS, or severe chronic airflow limitation, because it is often impossible to predict whether the underlying disease will allow sufficient recovery for successful weaning from the ventilator. Published figures from various centres indicate that the mortality of these patients while ventilated is very high[7 22] and associated with substantial human and material costs.[23] A decision to withhold mechanical ventilation in these patients should be made by a senior physician who first takes into account the present wishes of the fully conscious patient or the known wishes of the unconscious patient. In the unconscious patient the views of the closest relatives and of the staff directly concerned should also be considered. Where there is doubt, it is ethically more acceptable to withdraw treatment later than to withhold it in the moment of crisis.[13 14] All possible measures should be used in these patients to postpone the need for ventilation—for example, the use of doxapram if there is inadequate respiratory drive, continuous positive airway pressure if the main problem is airway collapse, and nasal intermittent positive pressure ventilation, which may be well

Table 4 Critical values of physiological variables widely accepted as part of the criteria for administering mechanical ventilation to adult patients (normal ranges in parentheses)*

VENTILATORY MECHANICS		
Tidal volume (VT; ml/kg)	<3	(5–7)
Respiratory rate (breaths/min)	>35	(12–20)
Minute ventilation (l/min)	<3 or >20	(6–10)
Vital capacity (VC; ml/kg)	<10–15†	(65–75)
FEV_1 (ml/kg)	<10	
Max inspiratory pressure	> −20–25†	(−75– −100)
VD:VT ratio	>0·6	(0·25–0·4)
GAS EXCHANGE		
Pao_2		
Fio_2 = 0·6	<8 kPa	
$P_A - ao_2$		
Fio_2 = 1·0	>46–60†‡ kPa	3·3–8·6
$Paco_2$	>6·7–8†§ kPa	4·6–6·0
CIRCULATORY VARIABLES		
Cardiac output	<2	l/min *or*
Cardiac index	<1·2	l/min/m^2

*The values in this table summarise those appearing in the current publications.[5–7 19 20 22] Consequently they are only approximate and VT, VC, and FEV_1 are not given in relation to age and sex.
†Range of published values.[7 19 22]
‡After at least 10 minutes of continuous Fio_2 = 1·0.
§In patients without metabolic acidosis or chronic hypercapnia.
VD—dead space; Fio_2—fractional inspired oxygen; Pao_2—arterial oxygen tension; $Paco_2$—arterial carbon dioxide tension; $P_{(A-a)}O_2$—alveolar-arterial oxygen difference.

tolerated for short periods, obviating the need for endotracheal intubation.[24]

It may be equally difficult to define the precise moment when ventilation should be started. Despite 70 years of worldwide experience with mechanical ventilation, there are no exact criteria on which to base a decision.[19] In the past 25 years, however, guidelines have evolved from a consensus of opinion among physicians and anaesthetists. Table 4 shows a set of physiological variables with a range of critical values that have been proposed by several authors. These critical values, however, are empirical, merely representing the accumulated experience of clinicians.[7 19 22] The decision to ventilate therefore rests firmly on the clinical findings. For example, having rapidly worsening respiratory variables is more important than exceeding a single critical value; fatigue and exhaustion cannot be quantified and should be judged by an experienced clinician. It is generally accepted, however, that reaching any of the critical values in table 4 is associated with terminal respiratory failure unless mechanical ventilation is in-

Table 5 Considerations in initiating or withholding ventilation

(*a*) *Clinical factors relevant to the decision to initiate or withhold ventilation in a patient with acute respiratory failure*
The conscious patient's acceptance of the treatment; degree of permanent mental impairment or other permanent, severe disability
History of previous admissions to intensive therapy unit and outcome of previous episodes of intermittent positive pressure ventilation
Is the underlying disease reversible?
Likelihood of successful weaning from the ventilator
Multiple organ failure?
(*b*) *Factors influencing the decision to initiate ventilation in acute respiratory failure*
Rapidly worsening physiological variables
Evidence of heart failure—fall in blood pressure, rise in heart rate, fall in urine output, etc
Presence of severe dyspnoea and sweating
Prominent use of accessory muscles, paradoxical movement of the abdomen
Inability to expectorate secretions
Severe fatigue of respiratory muscles—usually heralded by upward trends in respiratory rate and arterial carbon dioxide tension
Increasing confusion, restlessness, and exhaustion

stituted.[11] Table 5 lists the clinical factors which, complemented by the variables of table 4, influence the decision to start mechanical ventilation.

Ideally, severe respiratory failure should be anticipated and the decision to ventilate should be made before more than one of the critical values of table 4 is exceeded. Severe dyspnoea, restlessness, and exhaustion are in themselves good indicators for initiating ventilation when the underlying clinical condition is not expected to improve within one to two hours. Patients with impending acute respiratory failure need frequent and expert monitoring within an intensive care setting so that the decision to ventilate can be made at the appropriate time. Admission to the intensive therapy unit should therefore be arranged before mechanical ventilation is actually needed because a delay in the decision to ventilate may trigger a sequence of irreversible events, including multiple organ failure and cerebral oedema. Furthermore, the risks associated with sedation and endotracheal intubation increase as the clinical state of the patient deteriorates. The state of the patient is too often allowed to deteriorate too far in the general medical or surgical ward before being admitted to the intensive therapy unit. On the other hand, the usual consensus among physicians and surgeons, because of the high costs and the fierce competition for beds, is that admission to the intensive therapy unit is not warranted until intermittent positive pressure ventilation is essential. This deadlock, with obvious disadvantages to the patients, can be resolved only if clear policies for admission to the intensive therapy unit are established among the senior clinicians concerned. Scoring systems that help to predict the probability of survival of severely ill patients[7] may help in defining such admission policies.

High risk patients

Some clinical conditions deserve separate discussion because of their special risks and difficulties.

ACUTE ASTHMA OR ACUTE EXACERBATIONS OF CHRONIC OBSTRUCTIVE PULMONARY DISEASE

Respiratory failure is accompanied by chest hyperinflation, impaired diaphragm function, tachypnoea, and considerable use of accessory muscles. When intermittent positive pressure ventilation is applied inflation pressure may be very high (above 40 cm H_2O) with a very steep upward slope with each inflation, indicating very low chest compliance. The high intrathoracic pressures may have a substantial effect on the circulation, causing a fall in cardiac output. The risk of pneumothorax is high. When intermittent positive pressure ventilation is indicated in the exhausted, rapidly deteriorating patient it should not initially be aimed at correcting $Paco_2$; values up to 9 kPa are acceptable. Tidal volume should be small (under 0·6 l) and the ratio of inspiratory to expiratory time must be low (less than 1:4), allowing maximum time for the lungs to deflate in expiration[25]; a high oxygen concentration (at least 0·75) should be used to maintain oxygenation during the short period of life threatening respiratory failure. Saturated humidification of inspired gases, warmed to body temperature, is essential. The use of positive end expiratory pressure is contraindicated.

PATIENTS AT SPECIAL RISK FROM BAROTRAUMA

A proportion of patients with chronic respiratory failure have bullous emphysema, and intermittent positive pressure ventilation considerably increases the risk of pneumothorax. Positive end expiratory pressure is again contraindicated and inflation pressures should not exceed 40 cm H_2O with a maximum inspiratory:

expiratory ratio of 1:3, even if the patient remains underventilated (oxygenation should be ensured by raising FIO_2). The patient should be closely monitored for signs of tension pneumothorax (sudden increase in inflation pressure, tachycardia, and fall in blood pressure) and equipment should be ready for the insertion of a chest drain. When bullae burst and drains are inserted the problems of bronchopleural fistulas may supervene.

PATIENTS WITH STIFF LUNGS

Stiff lungs are usually the result of chronic interstitial disease or the adult respiratory distress syndrome secondary to cardiovascular shock, acid aspiration, or trauma. Intermittent positive pressure ventilation is used to replace the excessive work demanded of the respiratory muscles during the acute phase; high inflation pressures are needed, with positive end expiratory pressure if necessary to maintain oxygenation, so long as cardiac output is not severely compromised. In terms of peripheral oxygen delivery, a normal cardiac output with a PaO_2 of 6 kPa is better than half the normal cardiac output with a PaO_2 of 12 kPa. Low PaO_2 values are acceptable if the haemoglobin content of the blood is normal and cardiac output can reflexly rise above normal (haemodynamic monitoring and oxygen delivery is discussed in chapter 3). The high inflation pressures sometimes necessary in these patients lead to other problems, including pneumothorax and the difficulty of maintaining an adequate seal of the endotracheal tube without overinflating the cuff and damaging the trachea. It is impossible to predict accurately whether the respiratory muscles will be able to cope with chronically stiff lungs after the acute phase has passed.

LUNG SURGERY

Intermittent positive pressure ventilation should be avoided after lung surgery that includes bronchial resection because of the high incidence of bronchopleural fistula originating at the bronchial stump. If respiratory failure is due to lack of central drive doxapram could be tried before intermittent positive pressure ventilation is considered; narcotic analgesics should be avoided and pain relief by local analgesia is useful. If ventilation is necessary because of temporary respiratory failure endobronchial intubation of the intact side should be considered. If endobronchial intubation is not possible maximum inflation pressure, through a normal endotracheal tube, should be kept below 20 cm H_2O and positive end expiratory pressure should be avoided.

Special mechanical ventilation techniques

The "nuts and bolts" of ventilators and modalities of mechanical ventilation have already been reviewed (chapter 1). Intermittent positive pressure ventilation, with or without positive end expiratory pressure, is the modality of mechanical ventilation best suited for anaesthesia, postoperative ventilation, cardiopulmonary resuscitation, and most non-surgical conditions treated in the intensive therapy unit. Modifications of the original technique, which allow spontaneous or assisted breaths, are often used in the intensive therapy unit. The problems of weaning and ventilatory assistance are discussed in chapter 4. Two special techniques merit further discussion because of their potential application in difficult cases.

HIGH FREQUENCY VENTILATION

High frequency ventilation provides small inflations at rates of over 60 breaths/min (usually 200–300 breaths/min) by means of a special ventilator. The technique can maintain adequate gas exchange while achieving lower peak inflation pressures than conventional intermittent positive pressure ventilation. The mean airway pressure, however, is the same or higher than with the latter for equivalent alveolar ventilation, which reduces the possible advantages of high frequency ventilation to a few special circumstances.[26] There are unresolved problems with the equipment, such as humidification of inspired gases and monitoring of airway pressures during high frequency ventilation. It has been suggested that the technique has advantages over conventional intermittent positive pressure ventilation in the treatment of persistent bronchopleural fistula[27] and respiratory failure[28] associated with cardiac failure; both claims, however, have been disputed.[29,30]

DIFFERENTIAL VENTILATION

Differential ventilation allows ventilation of each lung with different gas mixtures or different pressure-time settings. It requires either a double lumen endotracheal tube or two cuffed bronchial tubes. Two synchronised ventilators are also needed. Experience with this technique is limited; the published evidence is based on case reports in patients with unilateral lung disease and on experimental animal work. Reported uses of differential ventilation include the treatment of patients with persistent bronchopleural fistula, unilateral bullae, and conditions in which copious secretions from one lung are affecting the function of the normal lung. If a high oxygen concentration or high inflation pressures are needed because of unilateral disease the "good" lung can be spared the risk of oxygen toxicity or excessive inflation pressures by differential ventilation.[31]

1 Drinker P, Shaw LA. An apparatus for the prolonged administration of artificial respiration. I. A design for adults and children. *J Clin Invest* 1929;7:229–47.
2 Lassen HCA. The epidemic of poliomyelitis in Copenhagen, 1952. *Proc R Soc Med* 1954;**47**: 67–71.
3 Ibsen B. The anaesthetist's viewpoint on the treatment of respiratory complications in poliomyelitis during the epidemic in Copenhagen, 1952. *Proc R Soc Med* 1954;**47**:72–4.
4 Björk VO, Engström CG. The treatment of ventilatory insufficiency after pulmonary resection with tracheostomy and prolonged artificial ventilation. *J Thorac Cardiovasc Surg* 1955;**30**:356–67.
5 Pontoppidan H, Geffin B, Lowenstein E. Acute respiratory failure in the adult. *N Engl J Med* 1972;**287**:690–8, 743–52, 799–806.
6 Petty TL, Lakshminarayan S, Sahn SA, Zwillich CW, Nett LM. Intensive respiratory care unit: review of ten years experience. *JAMA* 1975;**233**:34–7.
7 Knaus WA. Prognosis with mechanical ventilation: the influence of disease, severity of disease, age, and chronic health status on survival from an acute illness. *Am Rev Respir Dis* 1989;**140**:S8–13.
8 Zwillich CW, Pierson DJ, Creagh CE, Sutton FD, Schatz E, Petty TL. Complications of assisted ventilation: a prospective study of 354 consecutive episodes. *Am J Med* 1974; **57**:161–70.
9 Newman LH, McDonald JC, Wallace PGM, Ledingham IM. Propofol for sedation in intensive care. *Anaesthesia* 1987;**42**:929–37.
10 Kong KL, Willatts SM, Prys-Roberts C. Isoflurane compared with midazolam for sedation in the intensive care unit. *Br Med J* 1989;**298**:1277–80.
11 Cassidy SS, Eschenbacher WL, Robertson CH, Nixon JV, Bloomqvist G, Johnson RL. Cardiovascular effects of positive-pressure ventilation in normal subjects. *J Appl Physiol* 1979;**47**:453–61.
12 Luce JM. The cardiovascular effects of mechanical ventilation and positive end-expiratory pressure. *JAMA* 1984;**252**:807–11.
13 Wallis TW, Robotham JL, Compean R, Kindred MK. Mechanical heart-lung interaction with positive end-expiratory pressure. *J Appl Physiol* 1983;**54**:1039–47.
14 Cassidy SS, Mitchell JH. Effects of positive pressure breathing on right and left ventricular preload and afterload. *Fed Proc* 1981;**40**:2178–81.
15 Permutt S. Circulatory effects of weaning from mechanical ventilation: the importance of transdiaphragmatic pressure. *Anesthesiology* 1988;**69**:157–60.
16 Stinnett HO. Altered cardiovascular reflex responses during positive pressure breathing. *Fed Proc* 1981;**40**:2182–7.
17 Feuk U, Jakobson S, Norlén K. The effects of alpha adrenergic blockade on central haemodynamics and regional blood flows during positive pressure ventilation: an experimental study in the pig. *Acta Anaesthesiol Scand* 1987;**31**:748–55.
18 Kharasch ED, Yeo KT, Kenny MA, Buffington CH. Atrial natriuretic factor may mediate the renal effects of PEEP ventilation. *Anesthesiology* 1988;**69**:862–9.
19 Grum CM, Morganroth ML. Initiating mechanical ventilation. *J Intens Care Med* 1988;**3**: 6–20.
20 Snider GL. Historical perspective on mechanical ventilation: from simple life support system to ethical dilemma. *Am Rev Respir Dis* 1989;**140**:S2–7.

21 Pepe PE, Hudson LD, Carrico CJ. Early application of positive end-expiratory pressure in patients at risk for the adult respiratory distress syndrome. *N Engl J Med* 1984;**311**:281–6.
22 Hudson LD. Survival data in patients with acute and chronic lung disease requiring mechanical ventilation. *Am Rev Respir Dis* 1989;**140**:S19–24.
23 Rosen RL, Bone RC. Economics of mechanical ventilation. *Clin Chest Med* 1988;**9**:163–9.
24 Elliott MW, Steven MH, Phillips GD, Branthwaite MA. Non-invasive mechanical ventilation for acute respiratory failure. *Br Med J* 1990;**300**:358–60.
25 Tuxen DV, Lane S. The effects of ventilatory pattern on hyperinflation, airway pressures, and circulation in mechanical ventilation of patients with severe air-flow obstruction. *Am Rev Respir Dis* 1987;**136**:872–9.
26 MacIntyre N. New forms of mechanical ventilation in the adult. *Clin Chest Med* 1988;**9**: 47–54.
27 Gallagher TJ, Klain MM, Carlon GC. Present status of high frequency ventilation. *Crit Care Med* 1982;**10**:613–7.
28 Fuscardi J, Rouby JJ, Barakat T, Mal H, Godet G, Viars P. Hemodynamic effects of high-frequency jet ventilation in patients with and without circulatory shock. *Anesthesiology* 1986;**65**:485–91.
29 Bishop MJ, Benson MS, Sato P, Pierson DJ. Comparison of high-frequency jet ventilation with conventional mechanical ventilation for bronchopleural fistula. *Anesth Analg* 1987;**66**: 833–8.
30 Crimi G, Conti G, Bufi M, *et al.* High frequency jet ventilation (HFJV) has no better haemodynamic tolerance than controlled mechanical ventilation (CMV) in cardiogenic shock. *Int Care Med* 1988;**14**:359–63.
31 Carlon GC, Ray C, Klein R, Goldiner PL, Miodowwnik S. Criteria for selective positive end-expiratory pressure and independent synchronized ventilation of each lung. *Chest* 1978;**74**:501–7.

3

General care of the ventilated patient in the intensive care unit

MARTIN R HAMILTON-FARRELL, GILLIAN C HANSON

Care of the airway

Patients whose conscious level is impaired often require an artificial aid to maintain a clear airway. An oral (Guedel) airway may be sufficient temporarily, and it allows the passage of a suction catheter alongside. A nasopharyngeal airway is more comfortable, and may permit passage of a fine catheter through the larynx, though this may be traumatic if repeated too often.

An endotracheal tube or tracheostomy is necessary to secure the airway against laryngeal obstruction, to provide a route for artificial ventilation, to allow suction of bronchial secretions, and to protect the lungs against aspiration of pharyngeal and gastric contents (table 1). Early problems with endotracheal tubes include misplacement into the oesophagus or a mainstem bronchus;[1] and flexion, extension, or turning of the neck may displace the tube tip. Other complications include aspiration past an incompletely inflated balloon cuff, injuries of oropharyngeal mucus membranes, paralysis and granulomas of vocal cords, and laryngotracheal stenosis.[2] Accidental and self extubation is a risk, especially with young children,[3] and the inflated cuff may damage the larynx as it passes through (table 2).

Although the oral route is often easier for intubation, nasal

Table 1 Indications for endotracheal intubation

To obtain an airway secure from obstruction
To provide a route for artificial ventilation
To permit suction of bronchial secretion
To protect the lungs from aspiration of pharyngeal or gastric contents

endotracheal tubes have the advantages of avoiding trauma to the mouth and of being more comfortable for the awake patient. They are also longer and narrower than oral tubes, and occasionally present problems with suctioning. They may be associated with paranasal sinusitis,[4–7] particularly in patients receiving corticosteroids.

The use of plastic materials, standardised connector sizes, and uniform fixation techniques have helped to reduce some of the problems of endotracheal tubes.[8] The introduction of high volume, low pressure cuffs may help to prevent tracheal mucosal damage, though other factors are also important. Periods of hypotension and sepsis may compromise mucosal blood supply, and frequent changes of tube may damage the larynx. Tracheal mucosal damage is associated with local sepsis.[9]

Ulceration and granulomas of vocal cords occur in many patients after periods of intubation of five to fourteen days[10] and vocal cord paresis may also occur.[11] These complications are associated with laryngeal oedema, and usually resolve in the days or weeks after extubation. Tracheostomy will prevent these lesions.

Suctioning may cause mucosal damage, and ciliary action may be impaired by the frequent use of high vacuum suction apparatus attached to a catheter with a single end placed hole.[12 13] Pneumothorax may also occur with suctioning, especially in young children. Neonates are liable to develop subglottic oedema and stenosis after intubation.[14 15] Subsequent hoarseness and inspiratory stridor may be precipitated by an upper respiratory tract infection. In a small minority of cases, reparative surgery may be necessary.

Tracheostomy is necessary for patients requiring long term continuous positive pressure ventilation or continuous positive airway pressure, and those who need long term airway protection or tracheal suctioning.[16] Complications of intubation may also precipitate the decision to provide a tracheostomy (table 3). This is

Table 2 Complications of endotracheal intubation

Early
Misplacement into oesophagus or mainstem bronchus
Aspiration of pharyngeal or gastric contents past a deflated cuff
Oropharyngeal mucous membrane injuries

Late
Vocal cord paralysis and granulomata
Tracheal mucosal trauma from suctioning
Paranasal sinusitis (nasal tubes only)
Sepsis-bronchial or intrapulmonary

Table 3 Indications for tracheostomy

To relieve glottic/supraglottic obstruction
To facilitate long term ventilatory support
To facilitate continuous positive airway pressure (especially after chest injury)
To facilitate speech through artificial airway, using special adaptors

preferable to intubation for children, where the risk of self extubation requires constant observation—and a high level of staffing that is not always possible.[17] Once a tracheostomy has been provided, it offers the chance of staged decannulation, with the use of fenestrated tubes and tracheal buttons,[16] and this may permit an early transfer out of the intensive care unit.

The previous policy of carrying out a tracheostomy after a given period of intubation is becoming less popular.[18] Patients with facial injuries may require tracheostomy from the outset and patients who will clearly require intubation for a long time, such as those with the Guillain-Barré syndrome[19] and tetanus,[20] should receive a tracheostomy as soon as possible. Patients with chest trauma requiring continuous positive airway pressure may remain awake with regional analgesia, and find a tracheostomy more tolerable than intubation. On the other hand, with improvements in tube design and attention to risk factors for laryngeal damage, many consider that intubation may be tolerated for longer than was previously thought.[18]

Although the operating theatre provides the best surgical environment, there are risks associated with transferring patients from an intensive care unit. Tracheostomy can be performed in the intensive care unit so long as equipment and trained assistance are available for the surgeon.[21–23] A minitracheostomy is useful for patients who are unable to expectorate but who do not require continuous positive airway pressure or assisted positive pressure ventilation. It is used as an adjunct to regular physiotherapy.[24] Such procedures can be done at the bedside, with a local anaesthetic only.[25] But they carry many of the risks of formal tracheostomy,[22] and the potential for haemorrhage should not be underestimated.

Early complications of tracheostomy include pneumothorax and pneumomediastinum, subcutaneous emphysema, incisional haemorrhage, and tube displacement. Aspiration of gastric contents may occur during any airway manoeuvre where the protective reflexes are obtunded. The higher cricothyrotomy approach may produce laryngeal injury, but is not associated with pleural damage.

Table 4 Complications of tracheostomy

At insertion
Incisional haemorrhage
Pneumothorax, pneumomediastinum, subcutaneous emphysema
Early
Stomal haemorrhage
Aspiration of pharyngeal/gastric contents past a deflated cuff
Late
Stomal infection
Obstruction of lumen
Swallowing dysfunction
Erosion into oesophagus or innominate artery
Abnormal scar/granuloma formation
Tracheal stenosis at stoma or tip position
Sepsis-bronchial/intrapulmonary

Stomal bleeding is common, though usually not serious. Stomal infection is seen in about 12% of all tracheostomies,[23] and often arises on the fourth or fifth postoperative day (table 4).

Later complications of tracheostomy include tube obstruction, aspiration, swallowing dysfunction, and erosion into the oesophagus or the innominate artery. Even if stenosis does not occur, abnormal scar formation and granuloma may occur at the stoma site.[26] Tracheal stenosis produces stridor in adults only when over 75% of the lumen is obstructed.[27] The most direct method of assessing tracheal stenosis is by fibreoptic bronchoscopy or fibreoptic laryngoscopy, though this carries a risk from the anaesthetic. Lateral soft tissue radiographs of the neck with fluoroscopy are helpful. Maximum flow-volume loops and computed tomography are less sensitive methods of assessment.

All artificial airways require humidification or at least the conservation of exhaled water vapour. Heat-moisture exchangers are adequate for most cases, though high flow rates and tenacious secretions may require a heated humidifier.[28] The use of bacteriostatic materials is now common. Patients with tracheal tubes or tracheostomies are vulnerable to infection because of disrupted local clearance mechanisms, underlying immunosuppression, frequent suctioning, and the microbiological environment of the intensive care unit.[29] Viral infection is easily spread from staff to patients.[30] Contamination of the tracheal tube often precedes pneumonia by two to four days.[31] *Pseudomonas aeruginosa* is associated specifically with having a tracheostomy.[32]

Sedation and analgesia

Patients in the intensive care unit are exposed to many harmful stimuli. Some are later forgotten, but sensations that are even temporarily unpleasant should be avoided if possible. Among the common problems are anxiety, pain, lack of rest, thirst, tracheal tubes, face masks, nasogastric tubes, urinary catheters, and physiotherapy.[33] Any nursing procedure, such as turning or changes of dressing, are likely to be at the least uncomfortable.

Much can be achieved by careful explanation and appropriate reassurance from intensive care unit staff.[34] Attitudes to the use of sedative drugs have changed in the last few years[35,36] and there is increasing recognition that sedation, analgesia, and muscle relaxation should be provided specifically where indicated.[37–39] The use of muscle relaxants is restricted to patients in whom movement is dangerous (particularly with raised intracranial pressure, or where artificial ventilation is difficult, as in severe asthma). Muscle relaxants are not sedatives, and awareness of paralysis is terrifying and should be prevented by the use of sedatives.

Apart from subjective assessments of the adequacy of sedation, several sedation scoring systems have been designed. One system uses simple end points: fully alert, roused by voice, roused by pain, unrousable, paralysed, asleep (table 5).[40] Another identifies anxiety or restlessness as a separate category.[41] There are several other systems.[42,43] Such a system should be used by all staff in an intensive care unit, so that assessments can compare different patients and different drugs.

The simplest method of administering sedatives and analgesics uses repeated bolus doses. This, however, permits peaks and troughs of awareness and pain, which can be avoided by using intravenous infusions.[33] The institution of an infusion should be preceded by a loading dose, account being taken of any cardiovascular or respiratory depressant effects that the drug may have. The rectal route has also been used,[44] and subcutaneous infusions of

*Table 5 Sedation scoring system**

Fully alert
Roused by voice
Roused by pain
Unrousable
Paralysed
Asleep

*Shelly MP, Dodds P, Park GR.[40]

Table 6 The ideal sedative drug

Rapid onset and recovery after bolus or infusion
Wide therapeutic margin of safety
Minimal cardiovascular effects
Minimal and non-persistent respiratory depression
Water-solubility
Absence of metabolic, immunological, or hypersensitivity reactions
Absence of confusion after drug treatment is stopped
Low cost

opiates have been in use in our intensive care unit for several years.[45] Patient controlled sedation and analgesia are also used in some centres.[46]

The ideal sedative (and analgesic) drug has been described[47–49] as having the following properties: rapid onset and recovery by bolus or infusion; wide therapeutic index; minimal cardiovascular effects; respiratory depression that does not persist; water solubility; lack of irritation to veins; absence of metabolic, immunological, or hypersensitivity reactions; and absence of confusion after the drug is stopped. Such a drug should also be cheap. This is still a tall order, and no drug approaches it. Most intensive care units use a combination of opiates and benzodiazepines, given by bolus injection or infusion,[35] but there has never been a wholly satisfactory alternative to the effective but dangerous drugs etomidate and Althesin (table 6).[50]

Opiates are widely used, each intensive care unit having its current preference for one or another. All pure opiate agonists are respiratory depressants and antitussive agents, which may be useful; these properties, however, also delay weaning from the ventilator. Vasodilatation is common with opiates, which may be dangerous in the hypovolaemic patient; and gastric emptying and intestinal motility are slowed, which may delay enteral feeding.

Tolerance to analgesia is common, but addiction is very unlikely if the drug is used for a patient in pain.[51] Relying on opiates alone for sedation is inappropriate in view of their problems, especially in patients with renal impairment, where clearance may be very slow.[36 39]

Morphine is very useful (and cheap) but has active metabolites. Papaveretum, being half morphine, has similar properties, although its other constituent alkaloids may be more sedative. The short duration of a single bolus dose of fentanyl is due to redistribution rather than clearance, and its elimination half life is longer than that of morphine; there is therefore no indication for its use in intravenous infusions.[52]

Alfentanil,[43 47 53–58] with a short elimination half life and a small volume of distribution, is well suited to continuous intravenous use. It is dependent on hepatic metabolism, however, and is expensive. Pethidine has less sedative effect than the other opiates while being an effective analgesic, so it is useful in the awake and spontaneously breathing patient.

The use of regional analgesia is increasing. Epidural opiates do not require such precise placement of an epidural catheter as do local anaesthetics. The risk from respiratory depression is much reduced in an intensive care unit, where constant monitoring and observation should prevent complications. Non-opioid analgesics may be given intramuscularly or rectally. Indomethacin reduces opiate requirements, particularly after operation,[41] but may produce gastrointestinal bleeding.

The benzodiazepines are purely sedative, and used alone are inappropriate for patients in pain. They also reduce muscle tone, and promote amnesia. Diazepam has an active metabolite, *N*-desmethyl diazepam, which has a long elimination half life, and this reduces its usefulness for intravenous infusions.

Midazolam is water soluble and has an elimination half life of two to four hours after a single dose.[42] It is particularly useful for intravenous infusion but bolus doses may produce hypertension, especially in the hypovolaemic patient.[59] It has a large volume of distribution in some patients, particularly in those with hypoalbuminaemia.[60 61] This may explain the apparently unpredictable prolonged effect in patients with multiple organ failure.[62] Simultaneous infusion of a combination of alfentanil and midazolam has been recommended,[43 56] but the mixture of the two drugs in one syringe is therapeutically unsound as each patient has differing requirements for analgesia and sedation. The use of midazolam is common in children.[63 64]

Chlormethiazole given by intravenous infusion gives a rapid onset sedation that is easily adjusted, and it is specifically anticonvulsant[65]; it contains a large water load, however, and is slow to wear off.[66]

Chloral hydrate, often given to neonates and infants by nasogastrical tube, reduces opiate and benzodiazepine requirements.

Chlorpromazine, given intramuscularly or rectally, is useful for calming patients suffering from opiate or other drug withdrawal. Its alpha blocking properties may cause problems in hypovolaemic patients.

Ketamine, unlike other sedatives, increases arterial resistance and causes bronchodilatation. These effects may be useful in hypovolaemic and asthmatic patients, respectively. Concurrent use of midazolam may reduce psychological after effects.[67]

Propofol infusions are being used increasingly in the intensive care unit.[49 68–71] Our policy is to use this excellent but expensive agent for short periods only, particularly when the neurological state has to be examined frequently.

Inhaled sedative agents such as isoflurane have been introduced in some intensive care units,[72–74] where scavenging of exhaled gas is possible. The cost of isoflurane used in this way is high, and this has discouraged its widespread application.[75]

The pharmacological choice for sedation and analgesia in the intensive care unit is large but in the end each unit tends to have its own policy, which medical and nursing staff clearly understand. This is a sensible approach, so long as alternatives are constantly considered.

Non-respiratory monitoring of the ventilated patient

Haemodynamic monitoring is essential when the critically ill patient is ventilated. All patients should have a correctly sited central venous pressure (CVP) line. Central venous pressure measurements are useful in the fluid management of patients who have no appreciable pulmonary hypertension or cardiovascular disease. It is important for the patient not to have a low central venous pressure before assisted ventilation is started as the combination of an inadequate venous return to the right heart and positive pressure ventilation may lead to a catastrophic fall in blood pressure. The alterations in cardiac output during positive pressure ventilation have been ascribed to alterations in preload, with increased intrathoracic pressure causing peripheral translocation of central blood volume.[76] The elegant work of Wise *et al*[77 78] and Sylvester *et al*[79] has examined many of the factors affecting ventricular performance during mechanical ventilation with positive end expiratory pressure. It is now accepted that where respiratory reserve is limited and positive end expiratory pressure and ventilator settings have to be adjusted these adjustments should be made to optimise oxygen delivery and mixed venous oxygen saturation (which should if possible be over 70%). As with right atrial pressure, measurement of pulmonary capillary wedge pressure requires correct zero recordings and calibration; the measurement should be taken from the "a" wave at end expiration so that the level is not influenced by respiratory pressures.[80]

Other requirements for accurate measurement are enumerated in table 7. A pulmonary capillary wedge pressure below 18 mm Hg is not associated with pulmonary oedema related to pressure, pulmonary oedema below these pressures being related to permeability. Pressure related pulmonary oedema generally develops above pressures of 25 mm Hg, but in patients with longstanding high left atrial pressures oedema may not develop until the pulmonary capillary wedge pressure is over 30 mm Hg.

In patients with high airway pressures the difference between the pressure transmitted to the catheter tip (wedge pressure) and that transmitted to the left atrium cannot be predicted.[81] In critically ill patients the mean effective pulmonary capillary pressure (figure) may be a better measure than pulmonary capillary wedge pressure as any increase in the former may worsen gas exchange.[82]

Cardiac output measurements are usually made by using a pulmonary artery catheter with a thermistor probe at its tip. The thermodilution technique is as accurate and reproducible as the dye dilution technique, provided that the catheter is placed accurately, the volume and temperature (0°C) of the injectate are accurate, and the injection is always performed at the same point in the respiratory cycle. The derived variable, oxygen delivery, is of particular value for optimising ventilation and can be combined with the measurement of mixed venous oxygen saturation ($S\bar{v}O_2$) in the pulmonary artery, by blood sampling or the use of a catheter with an optical probe at its tip. Mixed venous oxygen saturation may not be an early predictor of change in cardiac output but it provides an indication of the delicate balance between oxygen delivery and extraction.

Table 7 Conditions for accurate measurement of PCWP in ventilated patients

Correct zero and calibration
Measure at end expiration from the 'a' wave
Ensure the catheter tip is sited below the level of the left atrium
Ensure no interference with pulmonary venous drainage between the catheter tip and the LA (for example, pulmonary embolus)
There should be no significant gradient between LA and LV (for example, MVD, where large 'v' waves may be mistaken for 'a' waves)
When the balloon is deflated, there should be a distinct PA trace; should a wedge trace appear, then the catheter must be withdrawn until the PA trace reappears

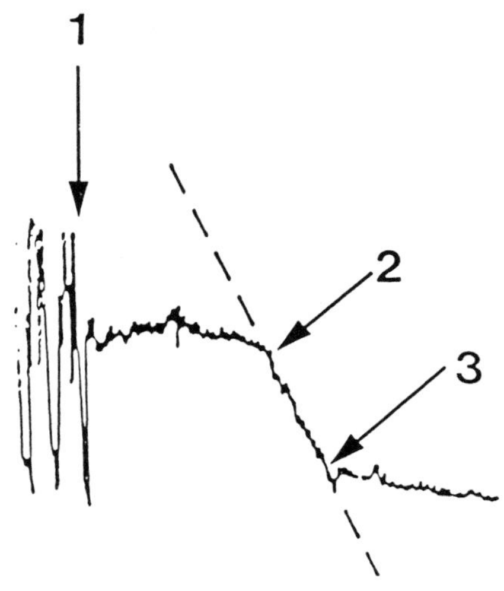

Mean effective pulmonary capillary wedge pressure (MEPCP)
1 System switched on for integrating pressures at 0·2 second intervals
2 Balloon inflation
3 MEPCP measured by a point at which a line drawn through the rapidly falling phase leaves the pressure trace

Indications for haemodynamic monitoring and measurement of derived variables are listed in table 8. Recent work has explained the importance of right ventricular performance in patients with acute respiratory failure[83] and how it changes with positive end expiratory pressure. Measurement of right ventricular volume variables has been made possible by the "fast response" thermodilution technique. Bronet and coworkers[83] found that, though in many patients with acute respiratory failure and increasing pulmonary hypertension the right ventricle dilated and right ventricular ejection fraction decreased, stroke volume was maintained unless there was concomitant disease, such as septic shock or viral myocarditis. Preload augmentation in these patients is clearly important, and it is

Table 8 Indications for measurement of cardiac output, oxygen delivery, and $S\bar{v}O_2$ in ventilated patients

Raised central venous pressure
Presence of pulmonary oedema on chest *x* ray
Hypotension in the presence of normal CVP and hypervolaemic response to volume challenge
Optimisation of haemodynamic variables in relation to ventilation
Optimisation of inotropic support

necessary in respiratory failure to ensure that left atrial pressures are maintained. Positive end expiratory pressure has been found to have two effects on right ventricular function.[84] In most patients it causes unloading of the right ventricle by reducing venous return, and in a few it leads to right ventricular dilatation and a decreased ejection fraction. A further study suggested that the changes in the right ventricle induced by positive end expiratory pressure are probably a function of the initial right ventricular ejection fraction and right ventricular end diastolic volume index.[85]

Nutrition in the ventilated patient

Acute weight loss of 30–40% of the original body weight is usually lethal,[86] and maximal physical performance is impaired in healthy individuals who have lost around 10% of body weight.[87] Increased mortality and morbidity correlate closely with an acute loss of body weight of 10–30% of the individual's normal weight.[88]

Clearly the best route for providing nutrition is the gastrointestinal tract. Calories given via the gastrointestinal tract maintain the integrity of the liver and oxidation of nutrients, a process that is imprecisely understood.[89] This route is often not suitable, however, in the ventilated patient because of the nature of the illness and the use of drugs that suppress bowel motility. When bowel absorption is not possible intravenous nutrition may be necessary, but it is important to reintroduce gastrointestinal feeding as soon as possible. When it is reintroduced after several days or weeks of parenteral nutrition, this must be done gradually; diarrhoea may ensue unless the osmolality of the feed is increased slowly over several days.

In the critically ill ventilated patient metabolic and volume normality must be established before nutrition is started (table 9). When a normal arterial carbon dioxide tension cannot be achieved by adjusting the ventilator settings, the total calories given as

Table 9 Checklist before feeding

All electrolyte deficiencies must be corrected
Acid base state must be stable
Adjust ventilation to ensure a normal $PaCO_2$
Ensure a normal calcium magnesium and inorganic phosphate concentration
Take a serum B_{12}, folate, and albumin before starting treatment
Cardiac and renal function must be known
Tolerance to a hyperosmolar load must be known
Is there evidence of hepatocellular failure?

glucose should be reduced to cut down carbon dioxide production.

Parenteral nutrition should be given via a designated feeding line. The line is best tunnelled subcutaneously and is generally inserted into the subclavian vein by the infraclavicular technique. This method may be complicated in the ventilated patient by a pneumothorax, and should therefore be avoided in patients with poor respiratory function (low arterial oxygen tension and saturation despite optimum ventilation and high fractional inspired oxygen), particularly when airway pressures are high. In these circumstances a feeding line may be inserted high into the internal jugular vein and tunnelled in the neck.

Energy requirements for the critically ill have traditionally been established by the use of the Harris-Benedict equation.[90] The total energy expenditure is then calculated by multiplying the basal metabolic rate by a stress factor, which may be augmented by an allowance for physical activity. This method was admirably described by Apelgren and Wilmore,[91] and is still being used in units where the total energy expenditure cannot be estimated by indirect calorimetry (table 10). Cortes and Nelson[92] showed that clinical assessment may overestimate energy expenditure because the

Table 10 Determination of energy requirements for TPN†

Energy requirements may be estimated by the following formula:
Normal (1) BMR × "Stress factor" (2) × 1·25 (3) = daily energy
Requirements for weight maintenance + 1000 (4) kcal − daily energy requirement for weight gain

1 Normal BMR can be determined by using standard nomograms or formulae (usually 1500 –1800 kcal/day)
Approximate values of resting metabolic rate for adults of average size are given below:

BW (kg)	50	53	60	65	70	75	80
kcal/day	1316	1411	1509	1602	1694	1784	1872

2 "Stress factor" is the normal BMR corrected for the disease process

Mild starvation	0·85–1·00
Postoperative (no complication)	1·00–1·05
Cancer*	1·10–1·45
Peritonitis*	1·05–1·25
Severe infection/multiple trauma*	1·30–1·55
Burn	1·50–1·70

3 The basal caloric requirements of the stressed patient are not adjusted upward when heavily sedated and ventilated, consider an increase of up to 20% in non-sedated patients

4 If anabolism and weight gain are the goals, an additional 1000 calories a day may be added

*Proportional to the extent of the disease
†Modified from Apelgren and Wilmore[91]

apparent degree of illness used as the basis for determining the stress factor is not an accurate guide. Not only is bedside calorimetry useful in more accurately assessing energy expenditure but it may also lead to financial saving. A nomogram for rapid calculation of metabolic requirements has been described.[93] It relies on the assumption that the respiratory quotient (RQ, the ratio of carbon dioxide production to oxygen consumption) is 0·8 and that expired gas can be analysed for carbon dioxide. Carbohydrate has an RQ of 1, fat 0·7, and protein 0·82, and lipogenesis (the synthesis of fat from glucose) has an RQ of 8. Carbohydrates given in excess of energy consumption promote lipogenesis, with a consequent increase in carbon dioxide production. In respiratory failure adjustment of the ventilator settings to lower the $PaCO_2$ may not be possible. In these circumstances, to avoid hypermetabolism related to a high glucose load, the proportion of carbohydrate to non-nitrogen containing calories should be 40–60%, the rest being given as fat.[94]

The use of fat rather than dextrose as a source of calories has two advantages. The respiratory quotient of fat oxidation is 0·7, compared with 1·0 with dextrose, so carbon dioxide production is lower; and fat, being iso-osmolar and of neutral pH, can be given via a peripheral vein.

The nitrogen requirements are generally assessed by measuring the urinary urea nitrogen loss in the urine.[95]

$$\text{Nitrogen (N) balance} = \underset{\text{IN}}{\text{grams of N}} - \underset{\text{OUT}}{\text{grams of urea N}} + 4\text{ g}$$

The 4 g factor accounts for the unmeasured nitrogen losses in skin and stool. Clearly, this method cannot be used to calculate nitrogen requirements when there are large losses from the bowel. Askanazi and coworkers[96] investigated the respiratory response of patients to increasing protein supply and found that progressive increase enhanced ventilatory drive and minute volume, but that in patients with limited reserve this may lead to respiratory failure in the spontaneously breathing patient. Precise protein requirements in patients requiring ventilatory support remain to be established.[96] Even in the presence of high nitrogen losses the nitrogen load is rarely increased above 18 g daily.

It is important in parenteral feeding to assess electrolyte and fluid balance regularly, with special attention to potassium and phosphate requirements. Hypophosphataemia is known to reduce oxygen transport and energy supply, and when severe may lead to respiratory failure.[97] Vitamins and trace elements should be replaced if feeding is continued for more than five days. Prolonged nutrition is best supplied by a mixture of protein, fat, electrolytes, vitamins, and trace elements made up into a 3 litre bag, prescribed early in the morning and supplied by the pharmacy later in the day. The mixture has to be carefully compiled to ensure compatibility between the different constituents of the feed.

Some studies suggest that ventilator weaning may be facilitated by preceding nutritional support.[98,99] Evidence suggests that respiratory muscle function is diminished in poorly nourished patients. Kelly[100] evaluated initial maximum inspiratory mouth pressure as an index of respiratory muscle function in 51 patients in hospital. Malnourished patients were found to have significantly less inspiratory force than normally nourished patients. The electrolyte content of the diet is particularly important during weaning—see table 11 (weaning is considered in more detail in the next chapter in this series). Further details on nutrition are supplied by Apelgren and Wilmore,[91] and, for acute respiratory failure, by Pingleton.[101]

Psychological and sleep disturbances during assisted ventilation

Psychiatric symptoms relating to assisted ventilation will not be manifest while the patient is heavily sedated and ventilation is controlled but are likely to emerge during weaning. It is clearly important to exclude organic dysfunction of the brain and physiological derangements due to blood gas or metabolic abnormalities.

The intensive care environment is recognised as causing stress, as a result of the alien and frightening atmosphere, sleep deprivation, unfamiliar noise, and a feeling of being confined by equipment, for

Table 11 Nutritional aspects during weaning

Observe $PaCO_2$ in relation to carbohydrate load
Ensure a non-protein calorie nitrogen ratio of around 150:1
Carbohydrate load should be 40–60% of total non-protein calories, the rest being given as fat
Nitrogen load should, in general, not exceed 14 g daily, should hyperventilation be present consider the possibility that nitrogen load is excessive
Ensure a normal serum phosphate and potassium
Observe fluid balance, do NOT overload

example.[102] Assisted ventilation produces additional stresses, related to awareness of the endotracheal or tracheostomy tube, the discomfort of suction and complicating hypoxia, and the horror of depending on a machine for ventilation.

Gries and Fernsler[103] conducted a survey to assess the causes of stress associated with ventilation in 17 patients, whose ages ranged from 35 to 81 years. Five patients could not recall the period of ventilation and three did not wish to discuss the problem. They categorised the stresses as shown in table 12. The major complaints were related to restriction of activity, the awareness and unpleasantness of the tracheal tube, suction, and the process of extubation. Inability to communicate was also frustrating. Some had vivid dreams, probably related to drugs. The more serious psychological problems—"the intensive care syndrome"—include disturbances of cognitive, affective, and perceptual functions. These occur in 12·5–18% of patients,[104–107] and are likely to be related to metabolic, neurological, or pharmacological factors. Many are related to drug dependence and withdrawal, and commonly occur during weaning. Hallucinations (pleasant and unpleasant) are common, and may manifest themselves as aggression, non-recognition of relatives, periods of agitation, and non-cooperation. These changes are most likely to arise in the patient who has been ventilated for a long time

*Table 12 Stressors associated with mechanical ventilation**

Intrapersonal, physiological
Frustration from activity restriction
Awareness of spontaneous breathing restriction related to ventilation
Intrapersonal, psychosocial, cultural
Insufficient explanation and hence misinterpretation of medical condition
Activity restriction, producing a feeling of
Ventilator dependence, inability to cope
Vivid dreams
Awareness of extubation
Interpersonal
Insufficient explanations
Inability to communicate
Loss of confidence in, and criticism of nursing care
Extrapersonal
Unpleasant experiences relating to
the tracheal tube
suctioning
extubation
noise—relating to the ventilator or surroundings

*Modified from Gries and Fernsler[103]

or who has previously been dependent on drugs—for example, on benzodiazepines, opiates, or alcohol. Drug withdrawal during weaning from the ventilator may be extremely difficult. Methadone is useful, enabling opiates to be rapidly reduced, and on occasion, where they are considered safe, beta blocking drugs may alleviate the tachycardia and anxiety related to opiate withdrawal. Rectal chlorpromazine given regularly, at the onset of weaning, has a potent calming effect without depressing the respiratory centre.

Gale and O'Shanick[108] discussed preventive psychological interactions for the ventilated patient, which are noted, with modifications, in table 13. Communication problems may in the future be reduced by the use of a word processor by the patient.[109]

Patients in an intensive care unit rarely sleep more than a few hours at a time,[110] and frequently do not complete a sleep cycle. Completion of a sleep cycle is essential for maintaining and restoring physical and psychological functions. Experiments in which people are deprived of sleep show that after two to five days subjects become anxious, suspicious, and disorientated, some

Table 13 Preventive psychological interactions

Communication problems
Talk to the patient
Provide alphabet, sign, or picture board
  writing tablet
  word processor

Dependence and loss of control
Allow the patient choice when possible (position in bed, radio station, etc)
Keep patient informed of progress
Inform patient of procedures to be undertaken, and the reasons for them

Fear of death or disability
Explain the ventilator and its alarm systems
Inform the patient of changes in ventilator settings and the reasons for them
Allow the patient to express his emotional problems and give support

Keep discussion and controversies about management away from the bedside

Isolation and fear of strangers
Establish continuity of care
Encourage visits by family and close friends

Sensory alteration
Maintain the patient's orientation with a calendar and clock on the wall, family photographs, and visits by relatives and close friends
Establish a day–night routine if at all possible
At night
  minimise noise
  minimise movement of and interference with the patient
  ensure adequate analgesia and maximise comfort
During the day provide a daytime environment with, for example: visits, television, music, etc.

developing delusions and paranoia.[111] Lack of prolonged sleep may be an important factor contributing to the intensive care syndrome.[112] Weissman and colleagues[113] found that the average length of a sleep period in an intensive care unit was only 24 minutes. The noise level in an intensive care unit is high, and has been found to be a major cause of sleep deprivation.[114 115] Methods whereby a normal sleep cycle can be encouraged are summarised in table 13.

The psychological needs of the ventilated patient have so far received scant attention. Aspects of care include maintaining a normal sleep cycle, minimising noise (particularly at night), awareness of the problems associated with drug dependence and withdrawal, helping and communicating with patients, and endeavouring to establish reorientation and a pleasant and calm environment during weaning.

1 Brunel W, Coleman DL, Schwartz DE, Peper E, Cohen HH. Assessment of routine chest roentgenograms and the physical examination to confirm endotracheal tube position. *Chest* 1989;**95**:1043–5.

2 Astrachan DI, Kirchner JC, Goodwin WJ Jr. Prolonged intubation vs. tracheostomy: complications, practical and psychological considerations. *Laryngoscope* 1988;**98**:1165–9.

3 Scott PH, Eigen H, Moye LA, Georgitis J, Laughlin JJ. Predictability and consequences of spontaneous extubation in a paediatric ICU. *Crit Care Med* 1985;**13**:228–32.

4 Aebert H, Hunefeld G, Regal G. Paranasal sinusitis and sepsis in ICU patients with nasotracheal intubation. *Intens Care Med* 1988;**1**:868–9.

5 Meyer P, Guerin JM, Habib Y, Levy C. Pseudomonas thoracic empyema secondary to nosocomial rhinosinusitis. *Eur Respir J* 1988;**1**:868–9.

6 Grindlinger GA, Neihoff J, Hughes SL, Humphrey MA, Simpson G. Acute paranasal sinusitis related to nasotracheal intubation of head-injured patients. *Crit Care Med* 1987; **15**:214–7.

7 Kronberg FG, Goodwin WJ Jr. Sinusitis in intensive care unit patients. *Laryngoscope* 1985; **95**:936–8.

8 Pippin LK, Short DH, Bowers JB. Long-term tracheal intubation practice in the United Kingdom. *Anaesthesia* 1983;**38**:791–5.

9 Chalon J, Ramanathan S. Care of the airway. *Int Anesthesiol Clin* 1986;**24**:53–64.

10 Kastanos N, Estopa-Miro R, Marin-Perez A, Xaubert-Mir A, Agusti-Vidal A. Laryngotracheal injury due to endotracheal intubation; incidence, evolution, and predisposing factors. A prospective long-term study. *Crit Care Med* 1983;**11**:362–7.

11 Marsh HM, Gillespie DJ, Baumgartner AE. Timing of tracheostomy in the critically ill patient. *Chest* 1989;**96**:190–3.

12 Loubser MD, Mahoney PJ, Milligan DW. Hazards of routine endotracheal suction in the neonatal unit. *Lancet* 1989;i:1444–5.

13 Bailey C, Kattwinkel J, Teja K, Buckley T. Shallow versus deep endotracheal suctioning in young rabbits: pathologic effects on the tracheobronchial wall. *Pediatrics* 1988;**98**:1165–9.

14 Dankle SK, Schuller DE, McClead RE. Prolonged intubation of neonates. *Arch Otolaryngol Head Neck Surg* 1987;**113**:841–3.

15 Jones R, Bodnar A, Roan Y, Johnson D. Subglottic stenosis in newborn intensive care unit graduates. *Am J Dis Child* 1981;**135**:367–8.

16 Stock MC, Woodward CG, Shapiro BA, Cane RD, Lewis V, Pecaro B. Perioperative complications of elective tracheostomy in critically ill patients. *Crit Care Med* 1986; **14**:861–3.

17 Arndal H, Andreassen UK. Acute epiglottitis in children and adults. Nasotracheal intubation, tracheostomy or careful observation? Current status in Scandinavia. *J Laryngol Otol* 1988;**102**:1012–6.
18 Watson CB. A survey of intubation practices in critical care medicine. *Ear Nose Throat J* 1983;**62**:494–501.
19 Gracey DR, McMichan JC, Divertie MB, Howard FM Jr. Respiratory failure in Guillain-Barré syndrome: a 6-year experience. *Mayo Clin Proc* 1982;**57**:742–6.
20 Udwardia FE, Lall A, Udwardia ZF, Sedhar M, Vora A. Tetanus and its complications: intensive care and management experience in 150 Indian patients. *Epidemiol Infect* 1987; **99**:675–84.
21 Hawkins ML, Burrus EP, Treat RC, Mansberger AR Jr. Tracheostomy in the intensive care unit: a safe alternative to the operating room. *South Med J* 1989;**82**:1096–8.
22 Stevens DJ, Howard DJ. Tracheostomy service for ITU patients. *Ann R Coll Surg Engl* 1988;**70**:241–2.
23 Dayal VS, el Masri W. Tracheostomy in intensive care setting. *Laryngoscope* 1986;**96**: 58–60.
24 Lewis GA, Hopkinson RB, Matthews HR. Minitracheostomy. A report on its use in intensive therapy. *Anaesthesia* 1986;**41**:931–5.
25 Hazard PB, Garrett HE Jr, Adams JW, Robbins ET, Aguillard RN. Bedside percutaneous tracheostomy: experience with 55 elective procedures. *Ann Thorac Surg* 1988;**46**:63–7.
26 Schloss MD, Gold JA, Rosales JK, Baxter JD. Acute epiglottitis: current management. *Laryngoscope* 1983;**93**:489–93.
27 Heffner JE, Miller KS, Sahn SA. Tracheostomy in the intensive care unit. Part 2: Complications. *Chest* 1986;**90**:430–6.
28 Cohen IL, Weinberg PF, Fein IA, Rowinski GS. Endotracheal tube occlusion associated with the use of heat and moisture exchangers in the intensive care unit. *Crit Care Med* 1988;**16**:277–9.
29 Salata RA, Lederman MM, Shlaes DM, *et al.* Diagnosis of nosocomial pneumonia in intubated, intensive care unit patients. *Am Rev Respir Dis* 1987;**135**:426–32.
30 Valenti WM, Clark TA, Hall CB, Menegus MA, Shapiro DL. Concurrent outbreaks of rhinovirus and respiratory syncital virus in an intensive care nursery: epidemiology and associated risk factors. *J Pediatr* 1982;**100**:722–6.
31 Comhaire A, Lamy M. Contamination rate of sterilized ventilators in an ICU. *Crit Care Med* 1981;**9**:546–58.
32 Freeman R, McPeake PK. Acquisition, spread, and control of *Pseudomonas aeruginosa* in a cardiothoracic intensive care unit. *Thorax* 1982;**37**:732–6.
33 Aitkenhead AR. Analgesia and sedation in intensive care. *Br J Anaesth* 1989;**63**:196–206.
34 Sedation in the intensive care unit [editorial]. *Lancet* 1984;i:1388–9.
35 Bion JF, Ledingham IM. Sedation in intensive care—a postal survey. *Intens Care Med* 1987;**13**:215–6.
36 Gast PH, Fisher A, Sear JW. Intensive care sedation now. *Lancet* 1984;ii:863–4.
37 Miller-Jones CM, Williams JH. Sedation for ventilation, a retrospective study of 50 patients. *Anaesthesia* 1980;**35**:1104–6.
38 Edbrooke DM, Hebron BS, Mather SJ, Dixon AM. Etomidate infusion: a method of sedation for the intensive care unit. *Anaesthesia* 1981;**36**:65.
39 Aitkenhead AR, Pepperman ML, Willatts SM, *et al.* Comparison of propofol and midazolam for sedation in critically ill patients. *Lancet* 1989;ii:704–9.
40 Shelly MP, Dodds P, Park GR. Assessing sedation. *Care of the critically ill.* 1986;**2**:170–1.
41 Ramsay MAE, Savage LTM, Simpson BRJ, Goodwin R. Controlled sedation with alphaxalone/alphadolone. *Br Med J* 1974;ii:656–9.
42 Dirksen MS, Vree TB, Driessen JJ. Clinical pharmacokinetics of long-term infusion of midazolam in critically ill patients—preliminary results. *Anaesth Intens Care* 1987;**15**: 440–4.
43 Hopkinson RB, O'Dea J. The combination alfentanil-midazolam by infusion: use for sedation in intensive therapy. *Eur J Anaesthesiol* 1987;**1**(suppl):67–70.
44 Reasbeck PG, Rice ML, Reasbeck JC. Double blind controlled trial of indomethacin as an adjunct to narcotic analgesia after major abdominal surgery. *Lancet* 1982;ii:115–8.
45 Waldman CS, Eason JR, Rambohun E, Hanson GC. Serum morphine levels. A comparison between continuous subcutaneous infusion and continuous intravenous infusion in post-operative patients. *Anaesthesia* 1984;**39**:768–71.
46 Loper KA, Ready LB, Brody M. Patient-controlled anxiolysis with midazolam. *Anaesth Analg* 1988;**67**:118–9.
47 Yate PM, Thomas D, Short SM, Sebel PS, Morton J. Comparison of infusion of alfentanil or pethidine for sedation of ventilated patients on the ITU. *Br J Anaesth* 1988;**61**:583–8.
48 Beller JP, Pottecher T, Lugnier A, Mangin P, Otteni JC. Prolonged sedation with propofol

in ICU patients: recovery and blood concentration changes during periodic interruptions in infusion. *Br J Anaesth* 1988;**61**:583–8.
49 Sear JW. Overview of drugs available for ITU sedation. *Eur J Anaesthesiol* 1987;**1**(suppl): 55–61.
50 Ledingham IM, Bion JF, Newman LH, McDonald JC, Wallace PG. Mortality and morbidity amongst sedated intensive care patients. *Resuscitation* 1988;**16**(suppl):S69–77.
51 Merriman HM. The techniques used to sedate ventilated patients: a survey of methods used in 34 ICUs in Great Britain. *Intens Care Med* 1981;7:217–24.
52 Shafer A, White PF, Schuttler J, Rosenthal MH. Use of a fentanyl infusion in the intensive care unit: tolerance to its anaesthetic effects? *Anaesthesiology* 1983;**59**:245–8.
53 Cohen AT, Kelly DR. Assessment of alfentanil by intravenous infusion as long-term sedation in intensive care. *Anaesthesia* 1987;**42**:545–8.
54 Sear JW, Fisher A, Summerfield RJ. Is alfentanil by infusion useful for sedation on the ITU? *Eur J Anaesthesiol* 1987;**1**(suppl):63–6.
55 Cohen AT. Experience with alfentanil infusion as an intensive care sedative analgesic. *Eur J Anaesthesiol* 1987;**1**(suppl):67–70.
56 Hoffman P. Continuous infusions of fentanyl and alfentanil in intensive care. *Eur J Anaesthesiol* 1987;**1**(suppl):71–5.
57 Sinclair ME, Sear JW, Summerfield RJ, Fisher A. Alfentanil infusions on the intensive therapy unit. *Intens Care Med* 1988;**14**:55–9.
58 Yate PM, Thomas D, Sebel PS. Alfentanil infusion for sedation and analgesia in intensive care. *Lancet* 1984;ii:396–7.
59 Geller E, Halpern P, Barzelai, *et al.* Midazolam infusion and the benzodiazepine antagonist flumezenil for sedation of intensive care patients. *Resuscitation* 1988;**16**(suppl):S31–9.
60 Shelly MP, Mendel L, Park GR. Failure of critically ill patients to metabolise midazolam. *Anaesthesia* 1987;**42**:619–26.
61 Vree TB, Shimoda M, Driessen JJ, *et al.* Decreased plasma albumin concentration results in increased volume of distribution and decreased elimination of midazolam in intensive care patients. *Clin Pharmacol Ther* 1989;**46**:537–44.
62 Bodenham A, Park GR. Reversal of prolonged sedation using flumazenil in critically ill patients. *Anaesthesia* 1989;**44**:603–5.
63 Silvasi DL, Rosen DA, Rosen KR. Continuous intravenous midazolam infusion for sedation in the paediatric intensive care unit. *Anesth Analg* 1988;**67**:286–8.
64 Shapiro JM, Westphal LM, White PF, Sladen RN, Rosenthal MH. Midazolam infusion for sedation in the intensive care unit: effect on adrenal function. *Anaesthesiology* 1986;**64**: 394–8.
65 Scott DB. Chlormethiazole in intensive care. *Acta Psychiatr Scand* 1986;**329**(suppl): 185–8.
66 Scott DB, Beamish ID, Hudson IN, Jostell KG. Prolonged infusion of chlormethiazole in intensive care. *Br J Anaesth* 1980;**52**:541–5.
67 Park GR, Manara AR, Mendel L, Bateman PE. Ketamine infusion. Its use as a sedative, inotrope and bronchodilator in critically ill patients. *Anaesthesia* 1987;**42**:980–3.
68 Farling PA, Johnston JR, Coppel DL. Propofol infusion for sedation of patients with head injury in intensive care. A preliminary report. *Anaesthesia* 1989;**44**:22–6.
69 Grounds RM, Lalor JM, Lumley J, Royston D, Morgan M. Propofol infusion for sedation in the intensive care unit: preliminary report. *Br Med J* 1987;**294**:397–400.
70 Aitkenhead AR. Propofol and intensive care. *Lancet* 1989;ii:1281.
71 Manners JM. Propofol and intensive care. *Lancet* 1989;ii:975.
72 Beechey AP, Hull JM, McLellan I, Atherley DW. Sedation with isoflurane. *Anaesthesia* 1988;**43**:419–20.
73 McLellan I. Isoflurane compared with midazolam in the intensive care unit. *Br Med J* 1989; **299**:259–60.
74 Kong KL, Willatts SM, Prys-Roberts C. Isoflurane compared with midazolam for sedation in the intensive care unit. *Br Med J* 1989;**298**:1277–80.
75 Park GR, Burns AM. Isoflurane compared with midazolam in the intensive care unit. *Br Med J* 1989;**298**:1642.
76 Seely RD. Dynamic effect of inspiration on the stroke volume of the right and left ventricles. *Am J Physiol* 1980;**154**:273–80.
77 Wise R, Robotham J, Bromberger-Barnes B, Permutt S. The effect of PEEP on left ventricular function in right heart bypassed dogs. *J Appl Physiol* 1981;**51**:541–6.
78 Wise R, Robotham J, Bromberger-Barnes B, *et al.* Elevation of left ventricle diastolic pressure by PEEP in the isolated in-situ heart. *Physiologist* 1979;**22**:134–8.
79 Sylvester JT, Goldberg HS, Permautt S. The role of the vasculature in the regulation of cardiac output. *Clin Chest Med* 1983;**4**:111–26.
80 Runciman WB, Putten AJ, Ilsley AH. An evaluation of blood pressure measurement.

Anaesth Intens Care 1981;**9**:314–25.
81 Downs JB, Douglas ME. Assessment of cardiac filling pressure during continuous positive pressure ventilation. *Crit Care Med* 1980;**8**:285–90.
82 Cope DK, Allison RC, Parmentier JL, Miller JN, Taylor AE. Measurement of effective pulmonary capillary pressure profile after pulmonary artery occlusion. *Crit Care Med* 1986;**14**:16–22.
83 Bronet F, Dhainaut JF, Devaux JY, Huyghebaert MF, Villemert D, Monsallier JF. Right ventricular performance in patients with acute respiratory failure. *Intens Care Med* 1988; **14**:474–7.
84 Neidhert PP, Suter PM. Changes of right ventricular function with positive end expiratory pressure (PEEP) in man. *Intens Care Med* 1988;**14**:471–3.
85 Brienza A, Dambrosio M, Bruno F, Marucci M,Belpiede G, Giuliani R. Right ventricular ejection fraction. Measurement in acute respiratory failure (ARF). Effects of PEEP. *Intens Care Med* 1988;**14**:478–82.
86 Moore FD. *Metabolic care of the surgical patient*. Philadelphia: WB Saunders, 1959:421.
87 Keys A, Brozek J, Henschel A, *et al. The biology of human starvation*. Vol 1. Minneapolis: University of Minnesota Press, 1950:714.
88 Studley HO. Percentage of weight loss. A basic indicator of surgical risk in patients with chronic peptic ulcer. *JAMA* 1936;**106**:458.
89 Defronzo RA, Jacot E, Jequier E, Maeder E, Wahren J, Felber JP. The effect of insulin on the disposal of intravenous glucose. *Diabetes* 1981;**30**:1000.
90 Harris JA, Benedict FG. *A biometric study of basal metabolism in men*. Washington DC: Carnegie Institute, publication No 279:1919.
91 Apelgren KN, Wilmore DW. Nutritional care of the critically ill patient. *Surg Clin North Am* 1983;**63**:487–507.
92 Cortes V, Nelson LD. Errors in estimating energy expenditure in critically ill surgical patients. *Arch Surg* 1989;**124**:287–90.
93 Smith HS, Kennedy DJ, Park GR. A nomogram for rapid calculation of metabolic requirements on intubated patients. *Intens Care Med* 1984;**10**:147–8.
94 Askanazi J, Nordenstrom J, Rosenbaum SH, *et al*. Nutrition for the patient with respiratory failure. Glucose vs fat. *Anaesthesiology* 1981;**54**:373–7.
95 Barrett TA, Robin AP, Armstrong MK, *et al*. Nutrition and respiratory failure. In: Bone RC, George RB, Hudston LD, eds *Acute respiratory failure*. New York: Churchill Livingstone, 1987:265–303.
96 Askanazi J, Weissman C, Lafala PA, Milic-Emili J, Kinney JM. Effect of protein intake on ventilatory drive. *Anaesthesiology* 1984;**60**:106–10.
97 Spector N. Nutritional support of the ventilator dependent patient. *Nursing Clin North Am* 1989;**24**:407–14.
98 Larca L, Greenbaum DM. Effectiveness of intensive nutritional regimens in patients who fail to wean from mechanical ventilation. *Crit Care Med* 1982;**10**:297–300.
99 Bassili HR, Deitel M. Effect of nutritional support on weaning patients off mechanical ventilation. *J Parenter Enter Nutr* 1981;**5**:161–3.
100 Kelly SM, Rosa A, Field S, Coughlin N, Schizgal HM, Macklem PT. Inspiratory muscle strength and body composition in patients receiving parenteral nutrition therapy. *Am Rev Respir Dis* 1984;**130**:33–7.
101 Pingleton SK. Nutrition in acute respiratory failure. *Lung* 1986;**164**:127–37.
102 Rappa D, Lavery R. Psychological dependences on mechanical ventilation resolved by "sigh-breathing" intermittent mandatory ventilation. *Respir Care* 1976;**21**:708–11.
103 Gries ML, Fernsler J. Patient perceptions of the mechanical ventilation experience. *Focus on Critical Care* 1988;**15**:52–9.
104 Kornfield DS, Zimberg S, Malm JE. Psychiatric complications of open-heart surgery. *N Engl J Med* 1965;**273**:287–92.
105 Lazarus LR, Hagens JH. Prevention of psychosis following open-heart surgery. *Am J Psychiatry* 1968;**124**:1190–5.
106 Holland J, Sgroi LSM, Marwit SJ, Solkoff N. The ICU syndrome: fact or fancy? *Psychiatry Med* 1973;**4**:241–9.
107 Kornfield DS, Heller SS, Frank KA, Moskowitz R. Personality and psychological factors in post-cardiotomy delirium. *Arch Gen Psychiatry* 1974;**31**:249–53.
108 Gale J, O'Shanick GJ. Pyschiatric aspects of respiratory treatment and pulmonary intensive care. *Adv Psychosom Med* 1985;**14**:93–108.
109 Cronin LR, Carrizosa AA. The computer as a communication device for ventilator and tracheostomy patients in the intensive care unit. *Critical Care Nurse* 1984; Jan–Feb:72–6.
110 Belitz J. Minimising the psychological complications of patients who require mechanical ventilation. *Critical Care Nurse* 1983; May–June:42–6.
111 Helton MC, Gordon SH, Nunnery SL. The correlation between sleep deprivation and the

intensive care unit syndrome. *Heart Lung* 1980;**9**:464–8.
112 Cousins MJ, Phillips GD. Sleep, pain and sedation. In: Shoemaker WC, Thomson WC, Holbrook PR, eds *Textbook of critical care*. Philadelphia: Saunders, 1984:797–800.
113 Weissman C, Kemper M, Elwyn DH, Askanazi J, Hyman AI, Kinney JM. The energy expenditure of the mechanically ventilated critically ill patient. *Chest* 1986;**89**:254–9.
114 Bentley S, Murphy F, Dudley H. Perceived noise in surgical wards and an intensive care area: an objective analysis. *Br Med J* 1977;ii:1503–6.
115 Redding JS, Hargest TS, Minsky SH. How noisy is intensive care? *Crit Care Med* 1977; **5**:275–6.

4

Weaning from mechanical ventilation

JOHN GOLDSTONE, JOHN MOXHAM

The capacity to ventilate mechanically the lungs has led to the widespread application of this technique in the intensive care unit, where the number of patients ventilated and surviving has been increasing since the early 1950s.[1] The indications for ventilatory support are now broad and include postoperative ventilation, cardiac failure, trauma, and ventilatory support in multiorgan failure in addition to ventilation for respiratory failure.[2] During recovery the transition from a positive pressure system (on the ventilator) to spontaneous, negative pressure breathing is in general accomplished without difficulty. Drugs are withdrawn, and the patient is allowed to make spontaneous efforts to breathe, either through the ventilator or from a simple breathing circuit. After a trial of unassisted breathing extubation usually follows, with or without supplemental oxygen.

The ability of a patient to breathe spontaneously after mechanical ventilation is related to many factors, including the diagnosis on admission and the length of time spent on the ventilator. In patients receiving short term ventilation as many as 20% of initial trials of spontaneous respiration may not be successful,[3] and further ventilation[4] or reintubation is required.[5] The incidence of weaning failure varies considerably, however; in a study of patients ventilated after cardiac surgery, where the period of elective ventilation had been a few hours, the overall incidence of initial failure to extubate was as low as 4%.[6]

Although 20% of patients ventilated acutely fail to be weaned initially, their progress and subsequent weaning is usually successful and rapid. Nett and coworkers showed that in such patients over

91% were able to breathe spontaneously after seven days.[7] In patients in whom weaning was still being attempted at one week the problems were complex. This group consists of patients with pre-existing lung disease as well as those patients surviving after severe multiorgan failure or neuromuscular disease, who tend to require ventilation for several days. Patients who have prolonged ventilation are more likely to require many days for weaning, and may take days or months to achieve spontaneous respiration by day and night.[8]

How then can we decide whether or not a patient is ready to be weaned successfully? The possibility of judging when a patient is able to breathe spontaneously has been examined largely in patients ventilated acutely, where investigators have documented various respiratory measurements before and during a trial of spontaneous respiration and compared the results with outcome (table 1). Unfortunately, there is little agreement in published reports about the power of these measurements to predict patients who will be unable to sustain spontaneous ventilation. For example, Tahvanainen and colleagues measured a battery of physiological parameters before extubation in a group of patients ventilated for either the adult respiratory distress syndrome, left ventricular failure, or neuromuscular disorders.[5] None of the conventional tests, including measurements of vital capacity, minute ventilation, respiratory rate, maximum voluntary ventilation (MVV), or maximum inspiratory pressure, identified patients that eventually required reventilation.

It has been suggested that simple clinical signs will detect patients who will fail to be weaned, including rapid shallow breathing[15] and respiratory paradox and alternans.[16] Although such clinical assessment is widely practised, published evidence to corroborate these signs is scarce. For example, in only about half of the studies is failure of weaning associated with increasing tachypnoea.[5 8 14 20 59] In

Table 1 Physiological measures conventionally associated with weaning failure

Tidal volume	< 5 ml/kg[9]
Vital capacity	< 10 ml/kg[10]
Minute ventilation (MV)	> 10 l/min[11 12]
Maximum voluntary ventilation	< 2 × MV[4 13]
Maximum inspiratory pressure	> −20 cm H_2O[4]
Alveolar-arterial oxygen tension difference	> 300 mm Hg[14]
Dead space/tidal volume	> 0·6[14]

Conversion to SI units: 1 mm Hg = 0·133 kPa; 1 cm H_2O = 0·098 kPa.

some patients, although the tachypnoea is a sign of eventual weaning failure, the high respiratory rate does not indicate the exact moment when reventilation is appropriate. In other studies there has been no difference between the respiratory rate in those who were weaned successfully and those who were not.[5 17 18]

The absence of a consensus on the value of signs and measurements predicting an individual patient's ability to be weaned successfully is reflected in clinical practice within the United Kingdom. In a survey of weaning practices in which 72% of 235 intensive care units in the United Kingdom responded the most common measures used were fractional inspired oxygen (FIO_2) and arterial oxygen tension (PaO_2); vital capacity was measured by only 35%, and maximum inspiratory pressure (PImax), maximum voluntary ventilation (MVV), compliance of the respiratory system, and the alveolar-arterial oxygen tension difference ($A\text{-}aDO_2$) were measured in less than a quarter of the units (J C Goldstone, unpublished findings). The deadspace-tidal volume ratio (V_D/V_T) and the pressure generated during the first 0·1 second of inspiration ($P_{0\cdot 1}$) were assessed by 6% of units. Most respondents stated that "clinical assessment," rather than tests, before and during periods of spontaneous breathing formed the basis for decisions on weaning.

Mechanisms of ventilatory failure

The ability to breathe spontaneously depends on three factors: central respiratory drive, the capacity of the respiratory muscles, and the load placed on the respiratory muscle pump. Hypercapnic respiratory failure will ensue when the balance between these factors is disrupted, either by a decrease in capacity (for example, in neuromuscular disease), an increase in load (for example, increased airway obstruction), or depression of central drive (for example, after a drug overdose). An approach towards answering the questions of when to withdraw mechanical ventilation and for how long and when to reinstitute support rests on the assessment of these three components—capacity, load, and drive.

CAPACITY OF THE RESPIRATORY MUSCLES IN THE INTENSIVE CARE UNIT

Measurement of the capacity of the respiratory muscles centres around the assessment of their ability to generate pressure. For the inspiratory muscles strength can be measured during a static effort against a closed airway, pressure being recorded at the mouth or in the endotracheal tube. Although methods may differ, most reports of maximum pressure generation in the intensive care unit show a large, up to 75% reduction in capacity.[4 19 20]

Before admission to the intensive care unit, the patient may have reduced strength of the respiratory muscles. Systemic disease may affect the respiratory muscles, at the level of the nerves,[21 22] the neuromuscular junction,[23] or the muscle itself.[24] This may exacerbate or precipitate respiratory failure. Pulmonary disease may adversely affect the mechanical performance of the respiratory muscles. With airways obstruction there is hyperinflation, muscle shortening, and a reduced capacity to generate inspiratory pressures. When low and flat the diaphragm is less effective at reducing pleural pressure and less able to raise gastric pressure and displace the abdominal contents to achieve a change in volume.

Respiratory muscle strength may diminish after admission to the intensive care unit (table 2). Metabolic abnormalities such as hypophosphataemia,[25] hypomagnesaemia,[26] and hypocalcaemia[27] may reduce muscle contractility acutely. The effect of hypoxaemia on muscle function is difficult to assess. Blood flow to muscle increases during hypoxaemia and may offset the decreased carriage of oxygen by blood, thereby maintaining oxygen delivery. In a carefully designed study, Ameredes *et al*[28] showed no change in muscle function during hypoxaemic conditions. Hypercapnia, however, decreases contractility,[29] especially if combined with acidosis. Hypoxia and hypercapnia may cause a synergistic decrease in force, as has been found in an animal model.[30]

Muscle performance may be diminished by infections. Venti-

Table 2 Factors that may impair respiratory muscle contractility in patients in the intensive care unit

Hypophosphataemia[25]
Hypomagnasaemia[26]
Hypocalcaemia[27]
Hypoxia
Hypercarbia[30]
Acidosis
Infection[31 32]
Disuse atrophy[34]
Malnutrition[35]

Table 3 Factors increasing the load on the respiratory muscles in patients in the intensive care unit

Bronchoconstriction[40]
Left ventricular failure[41]
Hyperinflation[52]
Intrinsic positive end expiratory pressure[52]
Artificial airways[47]
Ventilator circuits[51]

latory failure occurs as a result of respiratory muscle dysfunction in dogs with septicaemic shock.[31] During an upper respiratory tract infection muscle performance measured in terms of maximum inspiratory and expiratory mouth pressures is reduced by 30%.[32] Muscle atrophy occurs with disuse,[33] and this may be accelerated by sepsis. Anzueto *et al*[34] ventilated primates artificially and found after 11 days that diaphragm strength, measured during phrenic nerve stimulation, was reduced by 46%. Malnutrition occurs in many patients before admission to the intensive care unit, and may continue during the intercurrent illness. Respiratory muscle strength is reduced in undernourished patients[35] and the mass of the diaphragm is decreased in patients who are wasted.[36]

LOAD

During mechanical ventilation the work of breathing is performed by the ventilator, and is dissipated during gas compression, overcoming airflow resistance and inflating the chest against elastic components of the lung and chest wall. During spontaneous breathing work external to the lung is performed in moving gas in and out of the chest, which means overcoming the elastic forces of the lung and chest wall during inflation, the resistance to airflow, and minor forces of inertia and gravity. Not all work external to the lung can be measured, as some energy is expended during the breathing cycle that does not contribute to gas flow but deforms the chest wall. Although this may be substantial, the load applied to the respiratory muscles is largely related to the elastic and resistive elements during gas flow. In the intensive care unit the ventilatory load is often much higher than normal (table 3).

Load can be increased substantially by airways obstruction. During asthma induced by histamine challenge a fall in FEV_1 of 40% was associated with a threefold increase in load that required an eightfold increase in pressure generation per tidal breath.[37] In patients ventilated for left ventricular failure Rossi *et al*[38] measured compliance and airway resistance, and showed a substantial

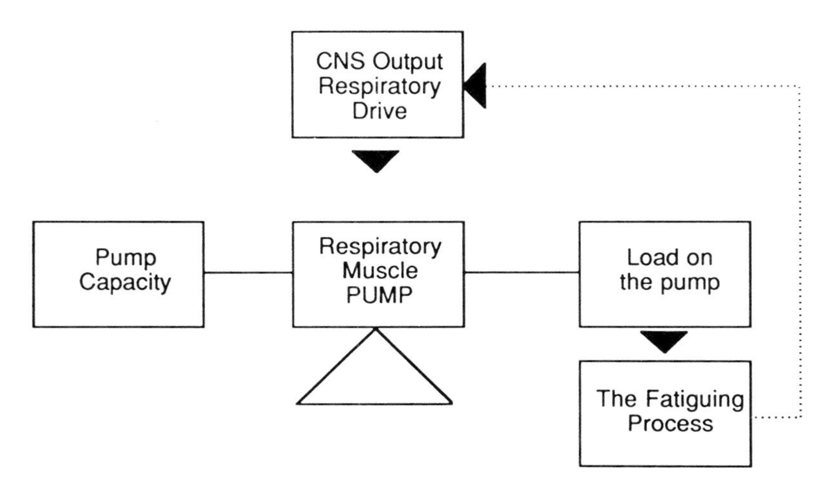

Diagram illustrating the central importance of the respiratory muscle pump and the crucial balance between load and capacity. When the ratio of load to capacity is high, the fatiguing process may be initiated, with (possible) adaptive changes in central nervous system respiratory drive.

increase in the load applied to the respiratory muscles. Resting oxygen consumption is increased in chronic airflow limitation, reflecting the increased work of breathing,[39] and in patients being weaned from ventilators the oxygen cost of breathing was four times greater than normal in patients with left ventricular failure.[40] Left ventricular function is impaired in many patients admitted to the intensive care unit, and pulmonary oedema increases the load substantially. This may occur during the transition to negative pressure breathing, as positive pressure ventilation may act to assist the left ventricle via transmitted pressure from the ventilator to the chambers of the heart.[41,42]

During weaning patients breathe through airways, apparatus, and ventilators, and this increases the load substantially.[43] The work required to breathe through an artificial airway is large,[44] greater than the work of breathing through the upper airway alone,[45] and it may double the load applied to the system.[46] The work needed to breathe through a tracheostomy may equal the work of breathing through the longer oral endotracheal tubes,[47] and may itself prevent spontaneous respiration.[48] Increased work is performed when patients are breathing through many circuits, especially when they are required to open valves to achieve inspiration.[49]

In many ventilated patients, especially those with airflow limitation, the time for expiration may not allow complete exhalation to functional residual capacity. Subsequent tidal breaths increase end expiratory volume and pressure; this is termed intrinsic or auto PEEP (positive end expiratory pressure). During a spontaneous breath the increased elastic recoil pressure of the lungs and chest wall must be overcome, and in patients who are weak this may be as great as half of their maximum inspiratory pressure generating capacity, which imposes a large additional load on the respiratory muscles.[50] Fiastro *et al*[51] measured the work of breathing during weaning from mechanical ventilation and found that patients able to breathe spontaneously had less work to perform than those who failed. In the "failed" group spontaneous respiration was achieved only when respiratory work was reduced to that observed in the successful group.

CENTRAL DRIVE

Force generation of the respiratory muscles is related to output from the central nervous system in terms both of the number of contractile units activated and of stimulation frequency. As motor neurone firing frequency is increased force increases rapidly, but it plateaus at frequencies greater than 50 Hz, with little increase at 100 Hz and beyond. In health and at rest low levels of central drive and concomitant low motor neurone firing frequencies are sufficient to effect an adequate tidal volume; patients with chronic respiratory failure have a higher respiratory drive,[52,53] placing them higher and less favourably on the frequency-force curve.

During weaning patients failing to achieve adequate ventilation have high central drive,[54] and indeed failure to breathe spontaneously has been correlated with an increased central drive that cannot be sustained.[55] Although occasional studies have shown that drive is reduced and may respond to central stimulants,[56] this has not been the case in most investigations. Central stimulants in patients breathing high on the frequency:force curve would not be expected to produce substantially greater ventilation. But any reduction in drive—due to sedation, for example—would lead to a large reduction in force generation, and to ventilatory failure.

THE BALANCE: RESPIRATORY MUSCLE FATIGUE

When the load applied to the respiratory muscles exceeds their capacity to generate pressure the likely outcome is the development of hypercapnic ventilatory failure, leading to acidosis, coma, and death. The hypothesis is that in these circumstances the respiratory muscles cannot sustain the required pressures without fatigue (figure).

Evidence supporting this hypothesis has largely come from studies in normal subjects breathing through inspiratory resistances. It has been shown that ventilation cannot be sustained when the pressure generated per breath exceeds 40% of maximum pressure.[57] The ability to maintain ventilation is also related to the duration of contraction of the inspiratory muscles during each breath. Bellemare and Grassino[58] performed repeated trials of inspiratory resistive loading measuring the time of inspiration (Ti) as a fraction of the respiratory cycle (Ttot). They defined a numerical relation between the strength of the diaphragm (Pdimax) and the duration and fraction of maximum pressure generated during each breath:

$$\text{Tension-Time Index} = \frac{\text{Pdi}}{\text{Pdi max}} \times \frac{\text{Ti}}{\text{Ttot}} .$$

This Tension-Time Index is about 0·05 during resting ventilation. When it exceeds 0·15, through an increase either in the duration of inspiration or in inspiratory pressure (induced experimentally by breathing through a resistance), ventilation cannot be sustained. Few studies have measured the Tension-Time Index in the intensive care unit during weaning. Pourriat and coworkers,[59] however, showed that patients who could not be weaned required a greater fraction of their maximum inspiratory pressure during each breath.

The relation between load, capacity, and fatigue in patients in the intensive care unit has been studied in terms of a modified Tension-Time Index, the Inspiratory Effort Quotient.[60] The mean pressure developed during inspiration is determined by tidal volume (VT) and dynamic compliance (Cdyn), and depends also on the shape of the inspiratory pressure curve (K):

$$\text{Inspiratory Effort Quotient} = \frac{(\text{k.V}_\text{T}/\text{Cdyn}) \times (\text{Ti}/\text{Ttot})}{\text{P}_\text{I}\text{max}} .$$

We measured the Inspiratory Effort Quotient in patients during trials of weaning and have found it to be low (0·05) in patients who were weaned satisfactorily and raised in those who could not be weaned and required ventilation.[61] This suggests that success with weaning is related to the balance between the strength of the respiratory muscles and the load applied to them rather than the absolute value of either measure. This may explain the variable predictive power in studies where aspects of strength or load are measured alone. Although the explanation of a high value for the Inspiratory Effort Quotient is complex in patients who cannot be weaned, high values for the Inspiratory Effort Quotient are likely to result in rapid weaning failure.

Given that excessive load in relation to capacity may lead to fatigue, a measurement of fatigue perhaps could be used to monitor patients being weaned from mechanical ventilation. This would enable patients to be reventilated at an appropriate moment, before the development of hypercapnia. With established fatigue maximum inspiratory pressures are reduced but measuring PImax in patients in the intensive care unit has hitherto been difficult and the results are variable. A recent approach in semiconscious patients is to measure the pressure generated by inspiratory gasps, when the patient is connected to a one way valve that is closed to inspiration. Pressure is measured at the peak of the inspiratory effort, after 8–10 occluded breaths or 20 seconds of occlusion.[62] This is a relatively simple technique, which can be used in patients not capable of making a maximum voluntary inspiratory effort and is surprisingly well tolerated in awake patients. Although this technique may allow a good measurement of PImax in stable patients it is unlikely to be easily applicable during the dynamic events of weaning.

Complex neuromuscular events occur during the fatiguing process initiated by the excessive loading of the respiratory muscles. In particular, the frequency distribution of the electromyogram changes during a fatiguing load.[63] The electromyogram of the diaphragm and other inspiratory muscles was shown to alter during weaning from mechanical ventilation in those patients who failed to breathe spontaneously,[16] but this technique is not in common use. Recording of electromyographic signals is difficult in patients in the intensive care unit and surface electrodes also record signals from non-respiratory muscles. During fatigue electromyographic changes occur early during loaded breathing, with little subsequent change when failure of force generation occurs, and they do not therefore indicate when reventilation should occur. Thus, although respiratory muscle fatigue probably occurs during weaning failure this phenomenon cannot be reliably detected with currently available techniques.

During muscular activity intense enough to lead to fatigue the speed of contraction and relaxation of muscle slows, and this can best be measured during the relaxation period after contraction has ceased.[64] During a brief inspiratory sniff the rate of relaxation of the respiratory muscles can be measured in terms of the maximum relaxation rate of oesophageal pressure.[65] In normal subjects during fatigue induced by loaded ventilation, the maximum relaxation rate slows and recovers rapidly with rest. Patients in the intensive care unit are usually intubated, with their upper airway bypassed, and unable to perform a sniff. A device that enables intubated patients to perform a sniff like manoeuvre has been used recently to study patients in the intensive care unit.[66] In those who could not be weaned the maximum relaxation rate slowed progressively and recovered after reventilation, suggesting that fatigue of the respiratory muscles was occurring during the attempt at weaning. Patients weaned successfully showed no slowing of the maximum relaxation rate of the respiratory muscles. In the future, maximum relaxation rate could perhaps be used as a reflection of the relation between the capacity of the respiratory muscles and the load that is applied to them when patients being weaned from mechanical ventilation are being assessed and monitored. Slowing of the maximum relaxation rate could provide an early indication that weaning will fail.

A strategy for weaning

The approach towards the patient who is likely, or has already proved, to be difficult to wean should begin by establishing a diagnosis or a list of clinical problems. When the causative factors that precipitated the need for ventilation are reversed, the patient may be a candidate for weaning. Much attention has been focused on the method of weaning patients from the ventilator, and opinion is divided. One method is to allow the patient to breathe spontaneously via a T piece circuit for gradually lengthening periods with full ventilation between these, and the other is to provide partial respiratory support by the ventilator and to allow the patient to breathe spontaneously between mechanical breaths (synchronous intermittent mandatory ventilation, SIMV—see chapter 2 for the different techniques of ventilation and support).[19] Currently there is no evidence to suggest that one method is superior to the other. Weaning is more likely to succeed in an alert, rested, cooperative patient. Sedation, confusion, and tiredness will make weaning less likely. In alert patients central respiratory drive is likely to be optimal; respiratory stimulants are of limited value and potentially harmful. At the centre of any weaning strategy is a detailed assessment of respiratory muscle capacity and load.

RESPIRATORY MUSCLE CAPACITY

In general, patients in the intensive care unit are weak, and small changes made to improve their strength or to reduce the load applied to the weakened respiratory muscles will be beneficial.

Correction of hypophosphataemia has been shown to increase strength and to facilitate weaning.[67] Electrolyte abnormalities should also be corrected. Although there is no direct evidence on the effect of hypercapnia and hypoxaemia during weaning, respiratory muscle function is likely to be reduced if the patient is acidotic, and tissue acidosis may be intensified by hypoxaemia. Nutritional support should be provided in the intensive care unit. Patients are often undernourished before admission to hospital, and the deficit may be large. Uptake of nutritional substrates may be impaired during episodes of critical illness, and intravenous feeding may be difficult in patients with complex problems of fluid balance. Patients can seldom be weaned during septic episodes, and weaning failure has been shown to be more likely in patients with a positive blood culture.[5] Respiratory muscle function may be diminished substantially by endotoxaemia. Although drugs, particularly aminophylline, have been reported to enhance respiratory muscle performance,[68] the balance of evidence suggests this is not the case.[69]

LOAD

During weaning the load applied to the muscles may alter acutely, precipitating respiratory failure and the need for reventilation.[51] Patients may have fluid overload or hypoalbuminaemia, leading to the development of pulmonary oedema at relatively low filling pressures. The mechanical enhancement of left ventricular performance by ventilation and the changes during weaning require consideration.

Airways obstruction increases the respiratory load and decreases respiratory muscle performance, and should be treated aggressively. Patients may be stable when assessed during mechanical ventilation yet may develop wheeze during spontaneous breathing, and should therefore be assessed during the weaning trial. Hyperinflation is likely in patients with airways obstruction, and this may be exacerbated by mechanical ventilation, which may increase intrinsic positive end expiratory pressure. Overdistension during mechanical ventilation can be monitored simply by displaying airway pressure during intermittent positive pressure ventilation on the bedside monitor, and watching for the characteristic waveform seen in such patients.[70] Intrinsic positive end expiratory pressure can be measured by occluding the expiratory limb of the ventilator during a prolonged expiratory pause, and measuring the airway pressure transmitted to the pressure gauge of the ventilator.[71] Patients susceptible to hyperinflation may breathe more effectively when removed from the ventilator altogether rather than having intermittent mandatory ventilation.[72]

Breathing apparatus may impose a substantial respiratory load on patients. Flow through endotracheal apparatus is affected by many factors, and is unlikely to be laminar in most cases. Resistance to flow increases with decreased tube diameter, and with a high minute ventilation may impose an unsustainable tension-time index of more than 0·15.[44] This load can be overcome by using inspiratory pressure support.

The benefit of positive end expiratory pressure (PEEP) is difficult to assess in the hyperinflated patient,[73] but it is of value in patients who have muscle weakness or obesity, or postoperative basal collapse. In such patients it increases functional residual capacity, prevents airways closure and atelectasis, increases compliance, and reduces ventilatory work. In these circumstances weaning is usually facilitated by adding continuous positive airway pressure (CPAP).

GENERAL MEASURES

Weaning, especially in patients who have been ventilated for many days or weeks, may be a great burden both physically and mentally. Sleep may be lost and disrupted and morale low, especially if the patient feels "stuck" on the ventilator. Although daytime respiratory drive should not be depressed, the establishment of regular sleeping patterns may require short acting sedative drugs.

The endpoint of a weaning trial is difficult to assess in some patients, as there are no current guidelines about the point where reventilation is mandatory—though the development of hypercapnia and acidosis indicates that reventilation is necessary. In studies of high intensity workloads in skeletal muscle, biopsy material has shown necrosis[74] and such changes probably occur in respiratory muscles if these are sufficiently stressed. Damage to the respiratory muscles, especially in patients who have severe weakness, only impedes successful weaning. In addition, the psychological effect of allowing a patient to breathe to the point of exhaustion demoralises the patient and erodes previous progress, and is therefore counterproductive.

1 Snider GL. Historical perspective on mechanical ventilation: from simple life support system to ethical dilemma. *Am Rev Respir Dis* 1989;**140**:S2–7.
2 Braun NMT. Intermittent mechanical ventilation. *Clin Chest Med* 1988;**9**:153–62.
3 Hilberman M, Kamm B, Lamy M, Dietrich HP, Martz K, Osborn JJ. An analysis of potential physiological predictors of respiratory adequacy following cardiac surgery. *J Thorac Cardiovasc Surg* 1976;**71**:711–20.
4 Sahn SA, Lakshminarayan S. Bedside criteria for discontinuation of mechanical ventilation. *Chest* 1973;**63**:1002–5.
5 Tahvanainen J, Salmenpera M, Nikki P. Extubation criteria after weaning from intermittent mandatory ventilation and continuous positive airway pressure. *Crit Care Med* 1983; **11**:702–7.
6 Demling RH, Read T, Lind LJ, Flanagan HL. Incidence and morbidity of extubation failure in surgical intensive care patients. *Crit Care Med* 1988;**16**:573–7.
7 Nett LM, Morganroth M, Petty TL. Weaning from mechanical ventilation: a perspective and review of techniques. In: Bone RC, ed. *Critical care: a comprehensive approach.* Park Ridge, Illinois: American College of Chest Physicians, 1984:171–88.
8 Morganroth ML, Morganroth JL, Nett LM. Criteria for weaning from prolonged mechanical ventilation. *Arch Intern Med* 1984;**144**:1012–6.
9 Radford EP, Ferris BG, Kriete BC. Clinical use of a nomogram to estimate proper ventilation during artificial ventilation. *N Engl J Med* 1954;**251**:877–84.
10 Bendixen HH, Egbert LD, Hedley-White J. *Management of patients undergoing prolonged artificial ventilation.* St Louis: CV Mosby, 1965:149–50.
11 Aldrich TK, Karpel JP. Inspiratory muscle resistive training in respiratory failure. *Am Rev Respir Dis* 1985;**131**:461–2.
12 Aldrich TK, Uhrlass RM. Weaning from mechanical ventilation: successful use of modified inspiratory resistive training in muscular dystrophy. *Crit Care Med* 1987;**15**:247–9.
13 Stetson JB, ed. *Prolonged tracheal intubation.* Boston: Little, Brown and Co, 1970:767–79.
14 Pontoppidan H, Laver MB, Geffin B. Acute respiratory failure in the surgical patient. In: Welch CE, ed. *Advances in surgery.* Vol 4. Chicago: Year Book Medical Publishers, 1970: 163–254.
15 Tobin MJ, Peres W, Guenther SM, *et al.* The pattern of breathing during successful and unsuccessful trials of weaning from mechanical ventilation. *Am Rev Respir Dis* 1986; **134**:1111–8.
16 Cohen CA, Zagelbaum G, Gross D, Roussos C, Macklem PT. Clinical manifestations of inspiratory muscle fatigue. *Am J Med* 1982;**73**:308–16.
17 Millbern SM, Downs JB, Jumper LC, Modell JH. Evaluation of criteria for discontinuing mechanical ventilatory support. *Arch Surg* 1978;**13**:1441–3.
18 Swartz MA, Marino PL. Diaphragmatic strength during weaning from mechanical ventilation. *Chest* 1985;**88**:736–9.
19 Krieger BP, Ershowsky PF, Becker DA, Gazeroglu HB. Evaluation of conventional criteria for predicting successful weaning from mechanical ventilatory support in elderly patients. *Crit Care Med* 1989;**17**:858–61.
20 Kacmarek RM, Cycyk-Chapman MC, Young-Palazzo PJ, Romagnoli DM. Determination of maximum inspiratory pressure: A clinical study and literature review. *Respir Care* 1989; **34**:868–78.
21 Cooper CB, Trend PStJ, Wiles CM. Severe diaphragm weakness in multiple sclerosis. *Thorax* 1985;**40**:633–4.
22 Al-Shaikh B, Kinnear W, Higgenbottam TW, Smith HS, Sneerson JM, Wilkinson I. Motorneurone disease presenting as respiratory failure. *Br Med J* 1986;**292**:1325–6.
23 Mier A, Brophy C, Green M. Respiratory muscle function in myasthenia gravis. *Am Rev Respir Dis* 1988;**138**:867–73.
24 Braun NMT, Arora NS, Rochester DF. Respiratory muscle and pulmonary function in

polymyositis and other proximal myopathies. *Thorax* 1983;**38**:616–23.
25 Newman JH, Neff TA, Ziporin P. Acute respiratory failure associated with hypophosphatemia. *N Engl J Med* 1977;**296**:1101–3.
26 Dhingra S, Solven F, Wilson A, McCathy D. Hypomagnesemia and respiratory muscle power. *Am Rev Respir Dis* 1984;**129**:497–8.
27 Aubier M, Viires N, Piquet J, *et al.* Effects of hypocalcaemia on diaphragmatic strength generation. *J Appl Physiol* 1985;**58**:2054–61.
28 Ameredes BT, Clanton TL. Hyperoxia and moderate hypoxia fail to affect inspiratory muscle fatigue in humans. *J Appl Physiol* 1989;**66**:894–900.
29 Juan G, Calverley P, Talamo C, Schnader J, Roussos C. Effect of carbon dioxide on diaphragmatic function in human beings. *N Engl Med J* 1984;**310**:874–9.
30 Esau SA. Hypoxic, hypercapnic acidosis decreases tension and increases fatigue in hamster diaphragm muscle in vitro. *Am Rev Respir Dis* 1989;**139**:1410–7.
31 Hussain SNA, Simkus G, Roussos C. Respiratory muscle fatigue: a cause of ventilatory failure in septic shock. *J Appl Physiol* 1985;**58**:2033–40.
32 Mier-Jedrzejowicz A, Brophy C, Green M. Respiratory muscle weakness during upper respiratory tract infections. *Am Rev Respir Dis* 1988;**138**:5–7.
33 Musacchia XJ, Deavers DR, Meininger GA, Davis TP. A model for hypokinesia: effects on muscle atrophy in the rat. *J Appl Physiol* 1980;**48**:479–86.
34 Anzueto A, Tobin MJ, Moore G, *et al.* Effect of prolonged mechanical ventilation on diaphragmatic function: a preliminary study of a baboon model. *Am Rev Respir Dis* 1987; **135**:A201.
35 Arora NS, Rochester DF. Respiratory muscle strength and Maximum Voluntary Ventilation in undernourished patients. *Am Rev Respir Dis* 1982;**126**:5–8.
36 Arora NS, Rochester DF. Effect of body weight and muscularity on human diaphragm muscle mass, thickness and area. *J Appl Physiol* 1982;**52**:64–70.
37 Martin JG, Shore SA, Engel LA. Mechanical load and inspiratory muscle action during induced asthma. *Am Rev Respir Dis* 1983;**128**:455–60.
38 Rossi A, Poggi R, Manzin E, Broseghini C, Brandolese R. Early changes in respiratory mechanics in acute respiratory failure. In: Grassino A, Fracchia C, Rampulla C, Zocchi L, eds. *Respiratory muscles in COPD.* London: Springer, 1988:149–60.
39 Lanigan C, Moxham J, Ponte J. Effect of chronic airflow limitation on resting oxygen consumption. *Thorax* 1990;**45**:388–90.
40 Field S, Kelly SM, Macklem PT. The oxygen cost of breathing in patients with cardiorespiratory disease. *Am Rev Respir Dis* 1982;**126**:9–13.
41 Beach T, Millen G, Grenvik A. Haemodynamic response to discontinuance of mechanical ventilation. *Crit Care Med* 1973;**1**:85–90.
42 Robotham JL, Cherry D, Mitzner W, Rabson JL, Lixfield W, Bromberger-Barnea B. A re-evaluation of the hemodynamic consequences of intermittent positive pressure ventilation. *Crit Care Med* 1983;**11**:783–93.
43 Marini JJ. The role of the inspiratory circuit in the work of breathing during mechanical ventilation. *Respir Care* 1987;**32**:419–30.
44 Shapiro M, Wilson RK, Casar G, Bloom K, Teague RB. Work of breathing through different sized endotracheal tubes. *Crit Care Med* 1986;**14**:1028–31.
45 Habib MP. Physiological implications of artificial airways. *Chest* 1989;**96**:181–4.
46 Wright PE, Marini JJ, Bernard GR. In vitro versus in vivo comparison of endotracheal tube airflow resistance. *Am Rev Respir Dis* 1989;**140**:10–6.
47 Plost J, Cambell JC. The non-elastic work of breathing through endotracheal tubes of various sizes [abstract]. *Am Rev Respir Dis* 1984;**129**:A106.
48 Criner G, Make B, Celli B. Respiratory muscle dysfunction secondary to chronic tracheostomy tube placement. *Chest* 1987;**91**:139–41.
49 Gibney RTN, Wilson RS, Pontoppidan H. Comparison of work of breathing on high gas flow and demand valve continuous positive airway pressure systems. *Chest* 1982;**82**:692–5.
50 Kimball WR, Leith DE, Robins AG. Dynamic hyperinflation and ventilator dependence in chronic obstructive pulmonary disease. *Am Rev Respir Dis* 1982;**126**:991–5.
51 Fiastro JF, Habib MP, Shon BY, Cambell SC. Comparison of standard weaning parameters and the mechanical work of breathing in mechanically ventilated patients. *Chest* 1988; **94**:232–8.
52 Gribben HR, Gardiner IT, Heinz CJ, Gibson TJ, Pride NB. The role of impaired inspiratory muscle function in limiting ventilatory response to CO2 in chronic airflow limitation. *Clin Sci* 1983;**64**:487–95.
53 Murciano D, Aubier M, Bussi S, Derenne JP, Pariente R, Milic-Emili J. Comparison of esophageal, tracheal and occlusion pressure in patients with chronic obstructive pulmonary disease during acute respiratory failure. *Am Rev Respir Dis* 1982;**126**:837–41.
54 Herrera M, Blasco J, Venegas J, Barba R, Dublas A, Marquez E. Mouth occlusion pressure

(P0·1) in acute respiratory failure. *Intens Care Med* 1985;**11**:134–9.
55 Sassoon CSH, Te TT, Mahutte CK, Light RW. Airway occlusion pressure: an important indicator for successful weaning in patients with chronic obstructive pulmonary disease. *Am Rev Respir Dis* 1987;**135**:107–13.
56 Hoake RE, Saxon LA, Bander SJ, Hoake RJ. Depressed central respiratory drive causing weaning failure. *Chest* 1989;**95**:695–7.
57 Roussos CS, Macklem PT. Diaphragmatic fatigue in man. *J Appl Physiol* 1977;**43**:189–97.
58 Bellemare F, Grassino A. Effect of pressure and timing of contraction on human diaphragm fatigue. *J Appl Physiol* 1982;**53**:1190–5.
59 Pourriat JL, Lamberto C, Hoang PH, Fournier JL, Vasseur B. Diaphragmatic fatigue and breathing pattern during weaning from mechanical ventilation in COPD patients. *Chest* 1986;**90**:703–7.
60 Milic-Emili J. Is weaning an art or a science? *Am Rev Respir Dis* 1986;**134**:1107–8.
61 Goldstone JC, Allen K, Green M, Moxham J. Sequential measurement of the Inspiratory Effort Quotient during weaning [abstract]. *Eur Respir J* 1990;**3**(S10):343S.
62 Marini JJ, Smith TC, Lamb V. Estimation of inspiratory muscle strength in mechanically ventilated patients: measurement of maximum inspiratory pressure. *J Crit Care* 1988; **1**:32–8.
63 Kaiser E, Petersen I. Frequency analysis of action potentials during tetanic contractions. *Electroencephalogr Clin Neurophysiol* 1962;**14**:955–60.
64 Esau SA, Bellemare F, Grassino A, Permutt S, Roussos C, Pardy RL. Changes in relaxation rate with diaphragmatic fatigue in humans. *J Appl Physiol* 1983;**54**:1353–60.
65 Koulouris N, Vianna LG, Mulvey DH, Green M, Moxham J. Maximum relaxation rates of oesophageal, nose and mouth pressures during a sniff reflect inspiratory muscle fatigue. *Am Rev Respir Dis* 1989;**139**:1213–7.
66 Goldstone JC, Allen K, Mulvey D, *et al*. Respiratory muscle fatigue in patients weaning from mechanical ventilation [abstract]. *Am Rev Respir Dis* 1990;**141**:A370.
67 Agusti AGN, Torres A, Estopa R, Agusti-Vidal A. Hypophosphatemia as a cause of failed weaning: The importance of metabolic factors. *Crit Care Med* 1984;**12**:142–3.
68 Aubier M. Pharmacotherapy of respiratory muscles. *Clin Chest Med* 1988;**9**:311–24.
69 Moxham J. Aminophylline and the respiratory muscles; an alternative view. *Clin Chest Med* 1988;**9**:325–36.
70 Milic-Emili J, Ploysongsang Y. Respiratory mechanics in the adult respiratory distress syndrome. *Crit Care Clin* 1986;**2**:573–84.
71 Pepe PE, Marini JJ. Occult positive end-expiratory pressure in mechanically ventilated patients with airflow obstruction. *Am Rev Respir Dis* 1982;**126**:166–70.
72 Williams MH. IMV and weaning. *Chest* 1980;**78**:804.
73 Marini JJ. Should PEEP be used in airflow limitation? *Am Rev Respir Dis* 1989;**140**:1–3.
74 Vihko V, Salminen A, Rantamaki J. Exhaustive exercise, endurance training and acid hydrolase activity in skeletal muscle. *J Appl Physiol* 1979;**47**:43–50.

5

Non-invasive and domiciliary ventilation: negative pressure techniques

JOHN M SHNEERSON

Historical development

Negative pressure ventilation first came into use during the second half of the nineteenth century. It was recognised that air would be drawn into the lungs through the mouth and nose if a sub-atmospheric pressure could be developed around the thorax and abdomen. When the pressure around the chest wall returned to that of the ambient air, expiration occurred passively owing to the elastic recoil of the lungs and chest wall.

The chest and abdomen are enclosed in an airtight, rigid chamber in all types of negative pressure ventilator, but in most of the earlier designs the whole of the body up to the neck was also contained in the chamber. This had the advantage that chest wall expansion was not limited by contact with the sides of the negative pressure device and that only one airtight seal, that around the neck, was required. The first of these negative pressure ventilators to be of clinical value was that developed at Harvard University Medical School by Drinker, an engineer, in 1928.[1] He designed several modified versions of his tank ventilator or iron lung and these were widely used during the poliomyelitis epidemics of the next 30 years. Tank ventilators were produced that could be constructed rapidly when needed and small enough for children.[2] Most of the early designs were telescopic in that the patient was pulled in and out of the main chamber of the tank on the mattress. In the later designs the upper part of the chamber was hinged towards the foot end and opened upwards (alligator type).

Simpler, non-tank negative pressure ventilators were first developed around the end of the nineteenth century. The most successful of these was Eisenmenger's biomotor, a cuirass designed in 1904.[3] This was superseded by various negative pressure shells or cuirasses introduced from 1930 to 1960. These were not as effective as the tank ventilators or the modern individually moulded cuirasses, but because of their simplicity they became widely used, particularly during recovery from acute poliomyelitis.

In the 1950s the jacket (wrap or poncho) design of negative pressure ventilator was produced,[4] in which the properties of rigidity and imperviousness to air were separated into two structures. The rigidity was provided by an inner framework of metal or plastic and this was covered by an airtight anorak-like garment with seals around the neck, arms, and usually the waist. The jacket, like the tank and cuirass, was connected to a pump, which generated a negative pressure between it and the patient's chest wall.

All these types of negative pressure ventilator were used both in hospital and at home to treat acute and chronic ventilatory failure. Ironically, as substantial improvements in their design were being made intermittent positive pressure ventilation using a translaryngeal endotracheal tube or a tracheostomy was shown to be more successful in the poliomyelitis epidemics, such as that in Copenhagen in 1952.[5] The superiority of intermittent positive pressure ventilation was probably due to better protection of the airway from aspiration. Negative pressure ventilators rapidly fell out of favour. In the last decade, however, negative pressure techniques have been used once again for various conditions, particularly neuromuscular and skeletal disorders, and have been shown to have a place, particularly for long term nocturnal ventilation in the home.[6]

Tank ventilation

Most of the modern tank ventilators are constructed of aluminium, though some, such as the Portalung, are made of plastic and are therefore lighter. The patient's body rests on a mattress within the chamber and a head and neck rest is provided in most designs to ensure comfort and to prevent kinking and obstruction of the upper airway. Most designs have windows that allow some observation of the patient and portholes through which catheters and monitor leads can be passed. They also enable some physiotherapy to be carried out while the patient is in the chamber and procedures such as arterial blood gas sampling to be performed. In several of the models either the head or the feet can be raised and in the Kelleher design the whole tank can rotate through 180° so that postural drainage can be carried out without removing the patient from the ventilator.[7]

The older tank ventilators have a separate bellows pump with a large stroke volume. In some designs this is incorporated into the structure of the ventilator. Newer models have separate rotary pumps of sufficient capacity to evacuate the large volume of air from within the tank chamber.

Tank ventilators are effective, do not need to be constructed individually to fit each patient and require only one airtight seal, around the neck. These advantages are, however, balanced by the lack of access to the patient and their size, weight, and cost. In addition, they are available in few centres in the United Kingdom and considerable experience is required to settle an ill patient successfully in the ventilator. An airtight, comfortable neck seal may be difficult to achieve and patients often find it awkward to lie in one position on their back for long periods.

Tank ventilation, like other forms of negative pressure and nasal intermittent positive pressure ventilation, does not protect the airway. Aspiration of material from the pharynx into the trachea and bronchi may occur, but in practice this is uncommon except in neuromuscular disorders associated with abnormalities of the swallowing mechanism. Tank ventilators are capable of maintaining normal blood gas tensions even if there is little or no spontaneous respiratory effort. The tidal volume is linearly related to the peak negative pressure within the chamber.[8] A pressure of around 30 cm H_2O is usually required but up to 40 cm H_2O may be necessary.

The models of tank ventilators that are used most frequently in the United Kingdom are those previously produced by Cape Warwick. There are several models, of which the "alligator" is the most common. A small number of Kelleher rotating ventilators survive and a simplified portable model was introduced in 1986 (fig 1). A few Both tank ventilators are also in use. These were first produced in 1938 and constructed of laminated wood. Most of them were modified in the 1950s but they are less satisfactory than the Cape Warwick ventilators. The most satisfactory ventilator of the newer, lighter, and more portable design is the Lifecare Portalung. This polycarbonate tank ventilator is available in three sizes. It is probably as effective as the traditional designs and the current cost is approximately £5000.

Figure 1 Tank ventilator: (a) closed, showing collar and head rest; (b) open, showing mattress and patient control panel. (Reproduced by permission of Penlon Ltd.)

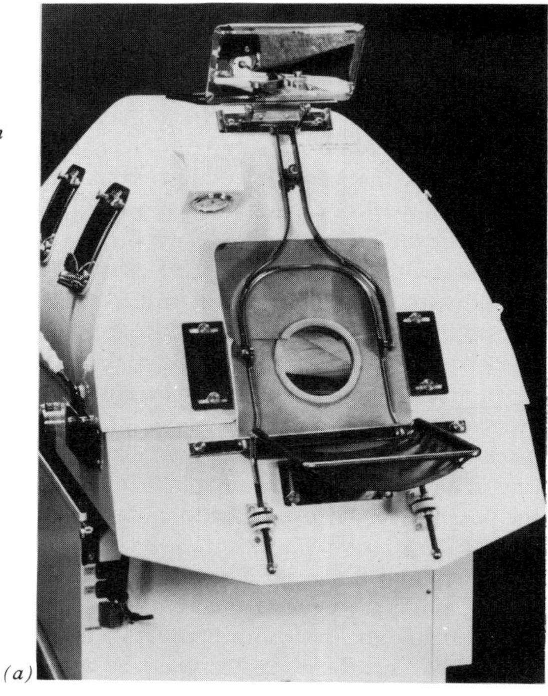

(a)

(b)

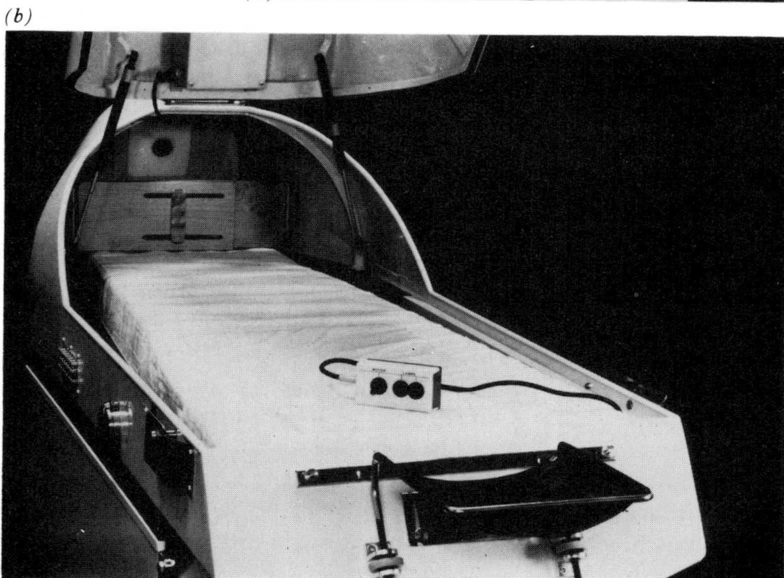

Jacket ventilation

The first effective jacket ventilator was the Tunnicliffe jacket, developed in the 1950s.[4] It is still produced in two sizes, but even the larger is too small for many patients, particularly those with scoliosis. Several newer models have been marketed in which the inner framework is made of a metal or plastic grid (fig 2). Some of these designs also have a back plate and they are all enclosed in an airtight synthetic garment. This may cover the legs as well as the trunk but in most designs the garment finishes below the hips. The air within the jacket is intermittently evacuated by a pump similar to that used for cuirass ventilation (see below).

The jackets do not restrain the expansion of the rib cage or abdomen, but they are awkward for many patients to put on and often cold to wear because of air leaks. Pressure areas are not a problem but jackets are invariably larger and more cumbersome than cuirasses. They are preferable to tank ventilators for home use but less satisfactory for treating the more severely ill patients in hospital. The tidal volume that they develop at any given pressure is less than that of a tank ventilator and the peak pressure that patients can tolerate is also usually slightly less.

Several jacket designs are currently available, costing about £500 plus the cost of the negative pressure pump. The most commonly used are the Tunnicliffe jacket, the Lifecare Pulmo-Wrap, and the Lifecare Nu-mo Garment.

Cuirass ventilation

Many of the earlier designs of cuirasses were ineffective because they were a poor fit and badly designed. Standard sized models rarely fit individual patients well enough both to be comfortable and to provide an airtight seal around the chest wall. Attention to detail in constructing the cuirass is important so that, for instance, the anterior abdominal wall is free to expand during inspiration and the movement of the lateral aspect and the upper chest is not restrained.

Standard sized designs of cuirass are still available but should seldom be used. Some suppliers will produce a cuirass from an individual patient's chest and abdominal dimensions, but these are rarely satisfactory.

It is usually best to construct the cuirass from a cast of the patient. This can be made of plaster of Paris, from which the cuirass is moulded.[9] The cuirass itself is constructed of light, airtight,

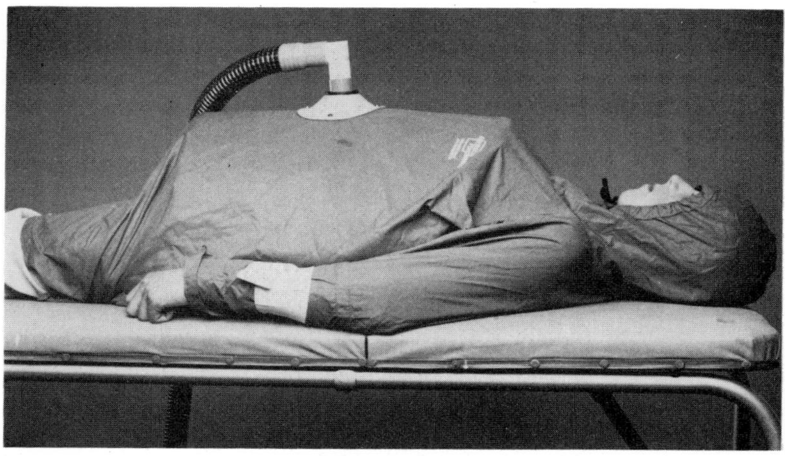

Figure 2 Jacket ventilator showing rigid internal framework surrounding thorax and abdomen, garment, and connecting tubing leading to negative pressure pump.

synthetic material, such as vitrathene or fibreglass. The edges are padded and covered with airtight material such as neoprene (fig 3). A back strap may be needed to maintain an airtight seal between the cuirass and patient.

Cuirasses can be made to fit patients even if there is a severe thoracic deformity. The construction of individually moulded cuirasses is analogous to the use of individually moulded nasal masks, which are widely used, particularly in Europe, for nasal intermittent positive pressure ventilation. The cost of the materials for a cuirass is less than £50.

Disadvantages of cuirass ventilation are that pressure areas may develop at the points of contact between the cuirass and the patient, and that if the patient grows or changes in weight a new cuirass may be needed. The patient has to sleep on his back or tilted slightly to one side and, particularly in patients with scoliosis, extra padding may be needed around the spine. Cuirasses are light and durable, and most patients can put them on without assistance.

The tidal volume achieved with a cuirass is linearly related to the peak negative pressure within it[10] and is considerably greater if the abdomen as well as the thorax is enclosed within the cuirass. Respiratory muscle activity is reduced[11] unless the patient is unable to coordinate with the pump.

Figure 3 Cuirass constructed of Vitrathene, with back strap attached and connector for tubing leading to negative pressure pump.

Various negative pressure pumps are available in the United Kingdom and are suitable for jacket or cuirass ventilation. The Cape cuirass pumps are still in use but are no longer manufactured. The most suitable are the Newmarket pump and the Lifecare 170-C pump. The Newmarket pump is a rotary pump and able to compensate for variable air leaks between the cuirass and the patient so that any preset pressure within the cuirass can be maintained.[12] The standard form provides controlled ventilation but there is also a patient triggered model.

Indications for negative pressure ventilation

ACUTE RESPIRATORY FAILURE

Neuromuscular and skeletal disorders Negative pressure ventilation may be invaluable in avoiding the need for endotracheal intubation during acute episodes of respiratory failure in these conditions.[13] Cuirass or jacket ventilation may be sufficient but many patients require treatment in a tank ventilator either continuously or at least overnight for a few days.

Any of the three types of negative pressure ventilator may be effective in weaning patients from translaryngeal positive pressure ventilation. Once a tracheostomy has been constructed, however, use of a cuirass is preferable to the other techniques as these may cause obstruction of the tracheostomy. After the acute episode has been effectively treated it is important to assess whether long term ventilatory support is needed and, if so, what form this should take.
Chronic lung disorders The place of negative pressure ventilation in acute infective exacerbations of chronic airflow obstruction has not been definitely established. There is some evidence that tank ventilation combined with energetic treatment of the underlying condition is effective and may avoid the need for endotracheal intubation.[14] Jacket and cuirass ventilators are much less effective in these circumstances. There have been no comparative studies of the value of tank and nasal positive pressure ventilation.

CHRONIC RESPIRATORY FAILURE
Neuromuscular and skeletal disorders These disorders are the main indication for negative pressure ventilation. Tank and jacket ventilators are effective but in most cases their inconvenience outweighs this advantage and a cuirass is preferable for long term use. Arterial blood gas tensions improve both during the day and at night,[10] and the prognosis once treatment is instituted is usually good unless the underlying disease is progressive.[15]
Chronic lung disease Recent studies have indicated that respiratory muscle strength[16] and endurance[17] may be increased by regular negative pressure ventilation. The evidence, however, is conflicting and the clinical importance of the physiological improvements that have been found is still uncertain.[18] There are no studies of the effectiveness of this type of treatment in improving prognosis but, as with intermittent positive pressure ventilation,[19] the results would seem likely to be no better than long term oxygen treatment.

Negative or positive pressure ventilation?

Many patients requiring mechanical respiratory support in the home need this only during sleep. When they are awake their respiratory drive is sufficient to maintain more or less normal blood gas tensions. In more mildly affected patients mechanical assistance may be required only during acute illnesses or postoperatively. Some patients with severe ventilatory failure, however, require mechanical support for their breathing for part of the day as well as each night.

There have been no comparative studies of cuirass and nasal positive pressure ventilation in these groups of patients. Less experience has been obtained with nasal intermittent positive pressure ventilation but comparison of the results of reported series suggests that the two methods are about equally effective. Upper airway obstruction may occur with all types of negative pressure ventilation, especially during rapid eye movement (REM) sleep. This may be due to inhibition of the upper airway abductor muscles as part of a generalised inhibition of inspiratory muscle activity, but loss of the normal sequence of upper and lower respiratory muscle activation may also be important. Obstruction is often reduced by lowering the peak negative pressure, or by giving the tricyclic antidepressant protryptyline, which reduces the duration of REM sleep. Nasal ventilation is preferable if upper airway obstruction during use of a cuirass cannot be overcome, and a cuirass is preferable if nasal intermittent positive pressure ventilation forces air through the cricopharyngeal sphincter into the stomach, causing abdominal distension. In most other patients either system is satisfactory and the choice depends on acceptability to the patient, local experience, availability of equipment, and cost. The capital cost of a cuirass and pump is slightly less than that of a nasal ventilator, and the servicing and maintenance are considerably cheaper. An assisted ventilation unit should have the facility to provide both types of treatment so that patients are not confined to either nasal intermittent positive pressure ventilation or negative pressure ventilation for lack of an alternative.

There is a smaller group of more severely affected patients, such as those with quadriplegia, who require continuous or almost continuous mechanical ventilatory support. In these circumstances positive pressure ventilation through a tracheostomy is usually required, though in selected cases alternative techniques, such as mouth intermittent positive pressure ventilation or phrenic nerve pacing,[2] may be of value for some or all of the time. Intermittent positive pressure ventilation through a tracheostomy with a cuffed endotracheal tube is also indicated if aspiration into the tracheobronchial tree is a problem owing to disordered swallowing, a poor cough, or both. Long term treatment in the home may be difficult to organise in these severely disabled and highly dependent patients.

Indications for long term domiciliary assisted ventilation

The decision to institute mechanical assistance to ventilation is an important one because this is usually a long term treatment. It is essential to establish that respiratory failure is present and that all conventional methods of treatment of the underlying cause have been tried. Unfortunately, there are no firm guidelines for judging whether respiratory failure is severe enough to require long term mechanical support. In general, however, this should be considered if respiratory failure has caused troublesome symptoms or potentially serious complications such as polycythaemia or pulmonary hypertension, or is likely to lead to these problems or to premature death.

Besides showing that the patient has severe respiratory failure, it is important to assess the probability of improving the outlook if treatment is instituted. In many neuromuscular and skeletal disorders the quality of life and the prognosis may be greatly improved but in chronic airflow obstruction there is no evidence yet that mechanical ventilatory support is more effective than long term oxygen treatment. In these patients ventilatory support does not prevent the slow and often inexorable progression of the underlying airway disease. The need to use a home ventilator reduces the quality of life and, as prognosis may not be improved, this treatment should be recommended only for carefully selected patients.

In some progressive neuromuscular disorders, such as Duchenne's muscular dystrophy, intermittent positive pressure ventilation through a tracheostomy may prolong life considerably but lead to a long period during which the quality of life is extremely poor because of severe and widespread muscle weakness. In these circumstances it may be considered ethical not to construct a tracheostomy for intermittent positive pressure ventilation but to provide a negative or nasal intermittent positive pressure ventilation system, neither of which protects the airway.

Funding and organisation of a home ventilator service

There are wide geographical differences in the use of domiciliary ventilatory support in the United Kingdom. These are due partly to variations in awareness of the value of treatment but also to the patchy availability of a home ventilator service.

The initiation of home ventilator treatment requires much more than simply providing suitable equipment and discharging the patient into the community.[20] A successful outcome requires explanation, education, and support for both the patient and the family or attendants. An understanding of the underlying cause of the respiratory failure and the equipment that has been provided, and what to do if problems arise, is needed. The choice of whether a negative or a positive pressure system is preferable often depends as much on the patient's personality and the degree of support in the home as on purely medical factors.

It is also essential that a rapid replacement service is available in case equipment fails. Within the supervising hospital there should be facilities and trained staff available continuously in case emergency readmission is needed.

At present there is no consistent pattern of NHS funding for domiciliary ventilatory support. Studies in the United Kingdom, France, and the United States have all shown that it is considerably cheaper than hospital care and just as safe.[19] The British government's proposals in the NHS and Community Care Act 1990 may rectify this once hospitals are reimbursed for the cost of treating individual patients. There is, however, the risk that the inevitable pressure to reduce costs will also reduce the standards of care and support, and limit the range of techniques available for treating this vulnerable group of patients.

Appendix: Suppliers of negative pressure equipment

TANK VENTILATORS

Portalung, available from Medicaid Ltd, Hook Lane, Pagham, Sussex PO21 3PP.

JACKETS

Tunnicliffe jacket, available from Watco Services (Basing Instruments Ltd), PO Box 86, Basingstoke, Hants RG24 0GZ.

Lifecare Pulmo-Wrap, available from Medicaid Ltd, Hook Lane, Pagham, Sussex PO21 3PP.

Lifecare Nu-mo Garment, available from Medicaid Ltd, Hook Lane, Pagham, Sussex PO21 3PP.

CUIRASS

Should be individually constructed.

PUMPS

Newmarket pump, available from Si-Plan Electronics Research Ltd, Avenue Farm Industrial Estate, Birmingham Road, Stratford-on-Avon, Warwicks CV37 0HP (about £2500).

Lifecare 170-C pump, available from Medicaid Ltd, Hook Lane, Pagham, Sussex PO21 3PP (about £5000).

1 Drinker PA, McKhann CF III. The iron lung. First practical means of respiratory support. *JAMA* 1985;**225**:1476–80.
2 Shneerson JM. *Disorders of ventilation*. Oxford: Blackwell, 1988:233.
3 Anonymous. Apparatus for maintaining artificial respiration. *Lancet* 1904;i:515.
4 Spalding JMK, Opie L. Artificial respiration with the Tunnicliffe breathing-jacket. *Lancet* 1958;i:613–5.
5 Lassen HCA. A preliminary report on the 1952 epidemic of poliomyelitis in Copenhagen with special reference to the treatment of acute respiratory insufficiency. *Lancet* 1953;i: 37–41.
6 Kinnear WJM, Shneerson JM. Assisted ventilation at home: is it worth considering? *Br J Dis Chest* 1985;**79**:313–15.
7 Kelleher WH. A new pattern of "iron lung" for the prevention and treatment of airway complications in paralytic disease. *Lancet* 1961;ii:1113–6.
8 Whittenberger JL, Ferris BG Jr. Alterations of respiratory function in poliomyelitis. *Am J Phys Med* 1952;**31**:226–37.
9 Brown L, Kinnear W, Sergeant K-A, Shneerson JM. Artificial ventilation by external negative pressure—a method for manufacturing cuirass shells. *Physiotherapy* 1985; **71**:181–3.
10 Kinnear W, Petch M, Taylor G, Shneerson JM. Assisted ventilation using cuirass respirators. *Eur Respir J* 1988;**1**:198–203.
11 Goldstein RS, Molotiu N, Skrastins R, Long S, Contreras M. Assisting ventilation in respiratory failure by negative pressure ventilation and by rocking bed. *Chest* 1987;**92**: 470–4.
12 Kinnear WJM, Shneerson JM. The Newmarket pump: a new suction pump for external negative pressure ventilation. *Thorax* 1985;**40**:677–81.
13 Libby DM, Briscoe WA, Boyce B, Smith JP. Acute respiratory failure in scoliosis or kyphosis. Prolonged survival and treatment. *Am J Med* 1982;**73**:532–8.
14 Lovejoy FW, Yu PNG, Nye RE, Joos HA, Simpson JH. Pulmonary hypertension. III. Physiologic studies in three cases of carbon dioxide narcosis treated by artificial respiration. *Am J Med* 1954;**16**:4–11.
15 Splaingard ML, Frates RC Jr, Jefferson LS, Rosen CL, Harrison GM. Home negative pressure ventilation: report of 20 years of experience in patients with neuromuscular disease. *Arch Phys Med Rehabil* 1985;**66**:239–42.
16 Scano G, Gigliotti F, Duranti R, Spinelli A, Gorini M, Schiavina M. Changes in ventilatory muscle function with negative pressure ventilation in patients with severe COPD. *Chest* 1990;**97**:322–7.
17 Cropp A, Dimarco AF. Effects of intermittent negative pressure ventilation on respiratory muscle function in patients with severe chronic obstructive pulmonary disease. *Am Rev Respir Dis* 1987;**135**:1056–61.
18 Zibrak JD, Hill NS, Federman EC, Kwa SL, O'Donnell C. Evaluation of intermittent long-term negative pressure ventilation in patients with severe chronic obstructive pulmonary disease. *Am Rev Respir Dis* 1988;**138**:1515–8.
19 Robert D, Gerard M, Leger P, *et al.* La ventilation mécanique à domicile definitive par tracheotomie de l'insuffisant respiratoire chronique. *Rev Fr Mal Respir* 1983;**11**:923–36.
20 Goldberg AI. Home care for life-supported persons: the French system of quality control, technology assessment and cost containment. *Publ Health Rep* 1989;**104**:329–35.

6

Non-invasive and domiciliary ventilation: positive pressure techniques

M A BRANTHWAITE

Intermittent positive pressure ventilation through an endotracheal or tracheostomy tube has been the mainstay of respiratory assistance for several decades. Some patients have been discharged home with a positive pressure device and tracheostomy but the number treated by these means in Britain has always been small.[1] Securing adequate gas exchange is usually easy and differs very little from that provided by the methods used in hospital (see chapter 2) but problems of organisation, finance, and training mean that this work has been concentrated in a few specialised centres where the necessary supervision and expertise were available.[2] These limitations have receded, however, with the advent of non-invasive methods of applying positive pressure to the airway, usually through a nasal mask but occasionally with a mouthpiece.[3] Other developments favouring the growth of domiciliary ventilation have been the recognition of nocturnal hypoventilation as a cause of respiratory and ultimately cardiorespiratory failure, the therapeutic success of mechanical ventilation used only during sleep, and a clearer understanding of the role of respiratory muscle weakness or fatigue. Assisted ventilation using negative pressure techniques is discussed in chapter 5; this chapter deals solely with non-invasive positive pressure methods.

Terminology and techniques

A fundamental distinction must be made at the outset between continuous positive airway pressure and intermittent positive pressure ventilation. The former refers to a spontaneously breathing patient in whom airway pressure is held positive in relation to atmospheric pressure throughout the respiratory cycle (fig 1). Intermittent positive pressure ventilation differs in that gas flow in and out of the lungs is controlled primarily by the ventilator, and the airway pressure changes phasically throughout the cycle. Confusion has arisen because a single design of mask can be used with both techniques. Some details of suitable positive pressure ventilators are provided in the table.

Nasal continuous positive airways pressure was introduced as a means of splinting the upper airway open during sleep in patients with obstructive sleep apnoea.[4] Pressures of 5–10 cm are used most often and values above 15 cm are rarely either needed or tolerated. The presence of a positive pressure in the airway tends to reduce the work of breathing and to increase functional residual capacity. Thus it can provide modest benefit to patients who do not suffer from obstructive sleep apnoea—for example, those with weak inspiratory muscles, who have a tendency to develop atelectasis. Some patients notice difficulty in expiration, however, particularly if there is a loss of lung or chest wall elasticity, so that active contraction of the expiratory muscles is necessary.

Intermittent positive pressure ventilation is intended to deliver the entire tidal volume if necessary and then allow passive exhalation, either to atmospheric pressure or to a predetermined positive airway pressure. A difference in airway pressure of 15–20 cm H_2O or more between inspiration and expiration is likely, whatever end expiratory pressure is selected. The technique can be used to control ventilation entirely or to augment spontaneous respiratory efforts. Augmenting gas exchange can be achieved

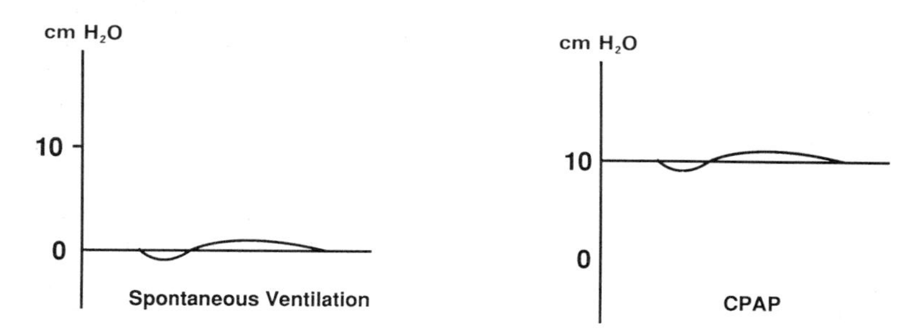

Figure 1 Diagrammatic representation of airway pressure profiles during spontaneous ventilation and continuous positive airway pressure (CPAP).

*Some ventilators suitable for nasal intermittent positive pressure ventilation that can be used in the home**

	Source	*UK agent or manufacturer†*	*Comments*
VOLUME PRESET VENTILATORS			
Bromptonpac/Airpac	UK	Pneupac Ltd, Luton	Independent compressor and ventilator
Lifecare PLV 100	USA	Medicaid Ltd, Bognor Regis	Integral battery
Monnal-D	France	Deva Medical Ltd, Runcorn	Maximum minute ventilation 20 litres on early model
Puritan-Bennett	USA	Puritan-Bennett UK, Hounslow	Several models with different capabilities
PRESSURE PRESET VENTILATORS			
Bantam	USA	Medicaid Ltd, Bognor Regis	Inexpensive; inflexible, no trigger
Respironics BiPAP	USA	Medicaid Ltd, Bognor Regis	Light weight; adjustable expiratory pressure; maximum positive pressure 28 cm; no disconnect alarm
Ventimate	UK	Thomas Respiratory Systems, London	Inexpensive; maximum pressure 50 cm

Respironics nasal masks are available from Medicaid Ltd and Sefam masks from Thomas Respiratory Systems.
*All are available in the United Kingdom; 1990 costs range from about £2500 to £7000.
†See appendix for addresses.

comfortably for conscious subjects only if the ventilator cycles into inspiration in response to the initiation of a spontaneous breath by the patient, a process described as "triggering" (fig 2). If there are no spontaneous inspiratory efforts or they are too feeble to trigger the ventilator, an automatic cycle must be imposed to ensure that gas exchange continues. The patient is then "controlled" by the ventilator. The so called assist/control mode can be used to ensure that breaths are triggered or imposed according to the magnitude of the spontaneous effort. It is important to recognise that when patients trigger the ventilator to deliver the next tidal volume they probably do not merely initiate the next breath. More often they continue to breathe throughout the delivery (fig 3) and this effort facilitates gas flow into the lungs and may enhance the total volume received. If on the other hand the start of the next breath is not followed immediately by gas flow from the ventilator there is a reflex increase in inspiratory effort, which is likely to make respiratory work greater than when they are breathing spontaneously. This means that the threshold of the ventilator trigger must be low and the response time short if the system is to be comfortable, easy to use, and capable of reducing respiratory work.

Triggering the ventilator occurs only if the spontaneous inspiratory effort is sufficient to reach a predetermined mechanical threshold, usually a reduction in airway pressure below atmospheric. Airflow limitation causing premature airway closure results in air trapping and hence a sustained positive pressure within the lungs at the end of expiration—known as endogenous or auto PEEP (positive end expiratory pressure).[5] This means that a considerable inspiratory effort is needed to lower the intrathoracic pressure to below atmospheric pressure so that a negative pressure sufficient to reach the trigger threshold is generated at the nose. This physiological constraint on the ease with which the ventilator can be triggered also exists if change in flow rather than pressure is the variable sensed at the nose. Spontaneous respiratory efforts conflict with the ventilator cycle if they fail to reach this threshold, though the resulting asynchrony is rarely perceived as discomfort.

A more subtle difficulty exists if a premature inspiratory effort reaches the trigger threshold too early in expiration, particularly if this occurs because the patient is breathing with the ventilator and continues to inspire after gas flow from the device has ceased. Successive breaths will "stack" and so cause hyperinflation because the machine is triggered into the next inspiratory phase before expiration has occurred to any appreciable extent. This can be prevented if there is a "closed window" in the respiratory cycle,

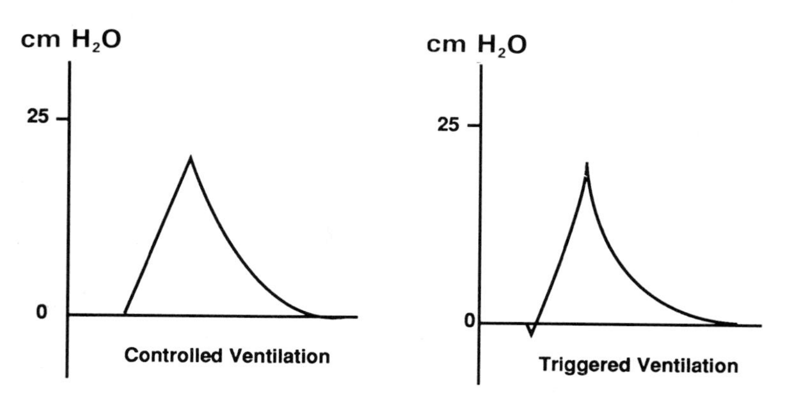

Figure 2 Diagrammatic representation of airway pressure profiles during controlled ventilation and triggered ventilation. Note the small negative pressure transient at the start of the triggered breath, which represents the patient's inspiratory effort.

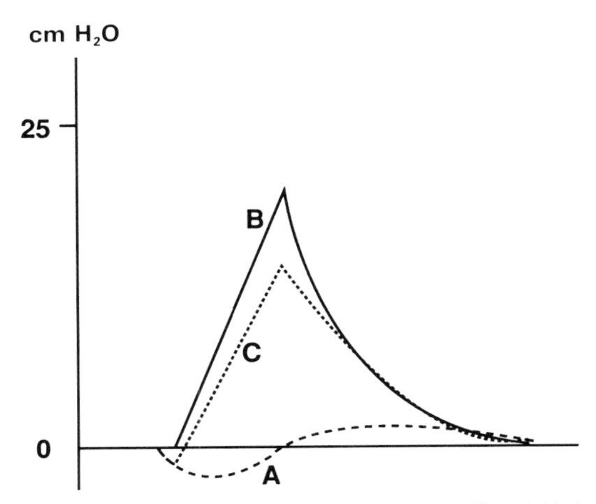

Figure 3 Diagrammatic representation of airway pressure profiles: (A) during spontaneous ventilation; (B) during fully controlled ventilation; (C) the resultant of the patient's inspiratory effort summating with the positive pressure pulse from the ventilator.

during which the trigger function is refractory. These requirements are needed particularly for patients who are conscious and unsedated or drifting off to sleep—when spontaneous breathing is often irregular in both depth and frequency. The alternative is to impose fully controlled ventilation, which some patients accept quite easily. They do so most readily if triggering the ventilator is difficult or uncomfortable, if their ventilatory requirements are easy to satisfy, and if they do not suffer from disorders that provoke dyspnoea independent of blood gas changes (for instance asthma, pulmonary oedema, and pulmonary fibrosis).

SELECTION OF PATIENTS [6]

Patients needing mechanical assistance to breathe for 16 hours or more each day are usually managed with a tracheostomy and positive pressure ventilator whether in hospital or at home. A chronic paralysing illness that does not interfere with mental function—for example, high transection of the spinal cord—is the usual kind of condition for which home care is considered. Phrenic pacing has a small role in these circumstances, usually as a means of enhancing daytime independence, but is feasible only if the nerve is capable of conduction and the diaphragm can contract effectively in response to the stimulus.

Non-invasive positive pressure ventilation is preferred for those who require ventilatory assistance only overnight. The hazards and discomforts of tracheostomy are avoided, external humidification is unnecessary, and daytime activities are totally unrestricted. Indications include central sleep apnoea, respiratory failure caused by thoracic deformity, static or only slowly progressive neuromuscular disease affecting the respiratory muscles, and patients with healed pulmonary tuberculosis treated by ablative measures that have resulted in a combination of obstruction and restriction. It is a matter for debate whether similar methods should be considered for patients with rapidly progressive neuromuscular disease affecting the respiratory muscles, even prophylactically, or for those with chronic obstructive pulmonary disease, in whom nocturnal exacerbation of hypoventilation is believed to contribute to daytime deterioration. Very few patients cannot be ventilated effectively with nasal intermittent pressure ventilation, but those unable to move their hands to their face will need to have the equipment fixed for them and may feel that it obtrudes on their limited ability to communicate. Negative pressure ventilation is sometimes preferred for these reasons.

TECHNIQUE OF NASAL POSITIVE PRESSURE VENTILATION [7]

Care is needed to ensure that the mask fits snugly and is comfortable. Several commercial models are available in multiple sizes, or individualised masks can be constructed to conform to facial contours. A poorly fitting mask is bound to leak. Efforts to secure it more firmly put pressure on the bridge of the nose, where the skin ulcerates very easily.

A tidal volume twice or one and a half times as great as that required during conventional intermittent positive posture ventilation is often needed. In part this reflects loss of gas through the open mouth or around the mask, but there is also more dead space because the entire upper respiratory tract is included in the ventilating volume.

The patient should be acclimatised to the system in hospital, where a period of intensive ventilatory assistance is often needed at the outset to restore normal daytime blood gas tensions and relieve heart failure. Intolerably breathless patients may have to be treated sitting upright but new users should preferably be trained while they are lying back on pillows or in an easy chair. This encourages them to relax and allows the operator to see whether accessory muscle activity has been abolished.

A few trial breaths should be offered first, asking the patient to hold the mask on his own nose, trying to prevent leaks around its edges and keeping the mouth closed firmly. Once confidence has been gained, the mask can be held in place for longer and longer periods while the operator adjusts the size and pattern of inspiration. It should be possible to abolish activity in the accessory muscles entirely, and tired or breathless patients often allow the ventilator to control them as soon as they feel they are receiving a sufficient volume. The relief of dyspnoea sometimes induces sleep almost immediately. Patients who are less ill at the outset may take longer to settle and their views on what settings are comfortable should be respected as far as possible.

The next stage is to secure the mask to the head with Velcro or elastic straps. This is best done with the mask disconnected from the ventilator. It can be reconnected as soon as the mask in in place and comfortable, but ventilation cannot be interrupted quite as quickly as when the mask is held by hand. It is worth starting with only two or three breaths before disconnecting the ventilator once more to check comfort and confidence.

Patients who are very breathless, severely hypoxaemic, or showing signs of circulatory distress should be given oxygen from the outset. Some ventilators provide for oxygen entrainment. Alternatively, a flow of 1 litre a minute, rarely more, added to a port on the front of the nasal mask usually suffices to achieve adequate arterial oxygenation and is unlikely to make an appreciable difference to either the settings on the ventilator or the resulting inflation pressure. Hypercapnia is less easily controlled if spontaneous respiratory efforts are abolished by relieving hypoxaemia but a rise in carbon dioxide tension is rare, even in patients who cannot tolerate any added oxygen while breathing spontaneously.[8]

Pulse oximetry provides a rough guide to safety initially but arterial blood gas measurements should be made half to one hour after the patient has been established on nasal intermittent positive pressure ventilation, when the carbon dioxide tension will have reached a plateau. Many patients, however, continue to contribute some spontaneous respiratory effort for several hours at least and so hypercapnia may return later unless the volume delivered from the machine is increased as they acclimatise to more fully controlled ventilation.

Patients should be encouraged to sleep as soon as possible while using the ventilator. There is no need to set the low pressure alarm (which detects leaks) with particular stringency at this stage, except for individuals particularly vulnerable to even a minor reduction in minute ventilation. Once the patient has acclimatised to sleeping for several hours or a whole night while using the machine, the alarm should be adjusted to ensure that appreciable leaks will be detected. Unfortunately not all patients are roused even by quite loud auditory signals. A pressure generating ventilator system, which provides some compensation for leaks, may be preferable in these circumstances.

The usual source of inadvertent volume loss is through the mouth in those who sleep with their mouths open. An elasticated chin-strap secured to the bands holding the mask to the head is often sufficient to control the leak but is rarely successful in the edentulous patient. The position is improved if dentures are retained during sleep, but this can cause soreness of the gums or jaws where the denture is pressed inwards and upwards by the mask and chin-strap.

Soreness on the bridge of the nose usually resolves if a soft wedge is used to lift the apex of the mask away from the face and the skin is protected with a light dressing. Occasional patients complain of either rhinorrhoea or excessive nasal dryness, sometimes accompanied by small epistaxes. Fortunately these symptoms are usually self limiting, although an external humidifier is sometimes necessary. Gaseous distension of the abdomen may be troublesome when the impedance to inflation of the chest is high or where controlled breaths are imposed, rather than delivered in response to the initiation of inspiration by the patient, when gas flow into the lungs is facilitated. Sometimes the delivered volume must be reduced to alleviate gastric distension, even though a lower inflation pressure and less effective gas exchange will result.

Few patients need to use the ventilator at home except overnight, and only a small minority with intrinsic lung disease or severe pulmonary hypertension require domiciliary oxygen as well. Special provision for emergency power supplies is not usually needed, and most patients can tolerate a night or two without support in the event of a mechanical failure.

The adequacy of overnight ventilation should be confirmed before the final arrangements for discharge home are made. Arterial blood gas tensions during mechanical ventilation by day are often misleading because the mouth is usually closed more firmly than during sleep and at least some spontaneous respiratory effort is likely to persist and augment gas exchange. Quite substantial reductions in ventilation may pass unnoticed if only oxygen saturation is recorded, and transcutaneous or end tidal carbon dioxide values should preferably be recorded as well.

Follow up

Outpatient follow up may be arranged as soon as good overnight control has been confirmed and the patient and family are familiar with the management and maintenance of the equipment. Regular measurement of arterial blood gas tensions is an essential component of follow up visits. Normal daytime values for arterial blood gas tensions are often restored and maintained when successful nocturnal ventilation is established for patients with respiratory failure caused by pure extrapulmonary restriction. More often the figures are not entirely normal, but an oxygen tension maintained above 8·0 kPa and a carbon dioxide tension below 7·0 kPa should be possible unless there is coincident lung disease. An exception is seen in patients with severe respiratory muscle weakness. No matter how good the overnight control, respiratory failure recurs by day because the spontaneously breathing patient cannot match ventilation achieved mechanically. This anomaly causes symptoms in a few cases, manifest as early morning dyspnoea when efforts to breathe spontaneously fail to match the preceding period of mechanical assistance.

The frequency of follow up should be determined by the nature of the disease, and factors such as the patient's motivation, compliance, and intelligence. An interval of three to six months is sufficient as soon as a satisfactory and stable condition has been achieved, and some patients need be seen no more than once a year. Reliable arrangements are necessary to ensure that contact can be re-established quickly if symptoms recur, usually as a result of respiratory tract infection. The same is true if any unrelated medical intervention is necessary, because the severity of the underlying respiratory disorder is often quite inapparent to others and undesirable risks may be incurred as a result. Regular maintenance of the equipment and provision of a prompt repair service are needed too; this should preferably be coordinated on a national basis in the interest of efficiency and economy.

Outcome

Domiciliary mechanical ventilation provides an impressive

improvement in longevity and quality of life in appropriate circumstances.[9,10] Patients and their families prefer treatment at home whenever possible, and this is not only desirable but also economically advantageous. Several countries have established a national system to provide comprehensive respiratory care for patients at home,[11,12] including the supervision of mechanical ventilation. It seems likely that similar developments will follow eventually in the United Kingdom.

Appendix: Addresses of manufacturers or UK agents mentioned in the table

Bromptonpac/Airpac:
Pneupac Ltd, Crescent Road, Luton LU2 0AH

Respironics BiPAP, Lifecare PLV 100, Bantam:
Medicaid Ltd, Hook Lane, Rose Green, Pagham, Bognor Regis PO21 3PP
(Respironics nasal masks also available from Medicaid)

Monnal D:
Deva Medical Electronics Ltd, 8 Jensen Court, Astmoor Industrial Estate, Runcorn WA7 1PF

Puritan-Bennett:
Puritan-Bennett (UK) Ltd, Heathrow Causeway, 152–176 Great West Road, Hounslow TW4 6JS

Ventimate:
Thomas Respiratory Systems, 33 Half Moon Lane, London SE24 9JX
(Sefam nasal masks are also available from Thomas Respiratory Systems.)

1 Dunkin LJ. Home ventilatory assistance. *Anaesthesia* 1983;**38**:644–9.
2 Spencer GT. Respiratory insufficiency in scoliosis: clinical management and home care. In: Zorab PA, ed. *Scoliosis*. London: Academic Press, 1977:315–28.
3 Braun NMT. Nocturnal ventilation—a new method. *Am Rev Respir Dis* 1987;**135**:523–4.
4 Sullivan CE, Issa FG, Berthon-Jones M, Eves L. Reversal of obstructive sleep apnoea by continuous positive airway pressure applied through the nares. *Lancet* 1981;i:862–5.
5 Marciano D, Aubier M, Bussi S, Derenne JP, Pariente R, Milic-Emil J. Comparison of esophageal, tracheal, and mouth occlusion pressure in patients with chronic obstructive pulmonary disease during acute respiratory failure. *Am Rev Respir Dis* 1982;**126**:837–41.
6 Kinnear WJM, Shneerson JM. Assisted ventilation at home: is it worth considering? *Br J Dis Chest* 1985;**79**:313–51.
7 Carroll N, Branthwaite MA. Intermittent positive pressure ventilation by nasal mask: technique and applications. *Intens Care Med* 1988;**14**:115–7.
8 Elliott MW, Steven MJ, Phillips GD, Branthwaite MA. Non-invasive mechanical ventilation for acute respiratory failure. *Br Med J* 1990;**300**:358–60.
9 Robert D, Gerard M, Leger P, *et al*. Domiciliary ventilation by tracheostomy for chronic respiratory failure. *Rev Fr Mal Respir* 1983;**11**:923–36.

10 Sawicka EH, Loh L, Branthwaite MA. Domiciliary ventilatory support: an analysis of outcome. *Thorax* 1983;**43**:31–5.
11 Goldberg AI. Home care for life-supported persons: is a national approach the answer? *Chest* 1986;**90**:744–8.
12 Peirson DJ. Home respiratory care in different countries. *Eur Respir J* 1989;**2**(suppl 7):630–5.

INDEX